149

Structure and Bonding

For further volumes:
http://www.springer.com/series/430

Aims and Scope

The series *Structure and Bonding* publishes critical reviews on topics of research concerned with chemical structure and bonding. The scope of the series spans the entire Periodic Table and addresses structure and bonding issues associated with all of the elements. It also focuses attention on new and developing areas of modern structural and theoretical chemistry such as nanostructures, molecular electronics, designed molecular solids, surfaces, metal clusters and supramolecular structures. Physical and spectroscopic techniques used to determine, examine and model structures fall within the purview of *Structure and Bonding* to the extent that the focus is on the scientific results obtained and not on specialist information concerning the techniques themselves. Issues associated with the development of bonding models and generalizations that illuminate the reactivity pathways and rates of chemical processes are also relevant.

The individual volumes in the series are thematic. The goal of each volume is to give the reader, whether at a university or in industry, a comprehensive overview of an area where new insights are emerging that are of interest to a larger scientific audience. Thus each review within the volume critically surveys one aspect of that topic and places it within the context of the volume as a whole. The most significant developments of the last 5 to 10 years should be presented using selected examples to illustrate the principles discussed. A description of the physical basis of the experimental techniques that have been used to provide the primary data may also be appropriate, if it has not been covered in detail elsewhere. The coverage need not be exhaustive in data, but should rather be conceptual, concentrating on the new principles being developed that will allow the reader, who is not a specialist in the area covered, to understand the data presented. Discussion of possible future research directions in the area is welcomed.

Review articles for the individual volumes are invited by the volume editors.

In references *Structure and Bonding* is abbreviated *Struct Bond* and is cited as a journal.

Mihai V. Putz • D. Michael P. Mingos
Editors

Applications of Density Functional Theory to Chemical Reactivity

With contributions by
A. Chatterjee • G. Maroulis • R.F. Nalewajski •
S. Pal • M.V. Putz

Springer

Editors
Mihai V. Putz
Structural and Computational Physical-Chemistry Laboratory
West University of Timisoara
Timişoara
Romania

D. Michael P. Mingos
Inorganic Chemistry Laboratory
University of Oxford
Oxford
United Kingdom

ISSN 0081-5993 ISSN 1616-8550 (electronic)
ISBN 978-3-642-44176-9 ISBN 978-3-642-32753-7 (eBook)
DOI 10.1007/978-3-642-32753-7
Springer Heidelberg New York Dordrecht London

Printed on acid-free paper

Springer is part of Springer Science+Business Media (www.springer.com)

Preface

For a long time, chemistry was considered as “ancilla Phisica” due to the commonly accepted hierarchy of scientific disciplines that implied a reduction in rigor from physics to chemistry. In the early twentieth century, following the elucidation of the structure of atoms, it became evident that atoms and molecules with even numbers of electrons were far more numerous than those with odd numbers of electrons. In 1916, G. N. Lewis provided the first comprehensive description of ionic and covalent bonds, when he postulated that atoms tend to hold an even number of electrons in their outer shells and a special stability was associated with eight valence electrons, which he speculated were arranged symmetrically at the eight corners of a cube. In 1919, I. Langmuir suggested that the structure of the periodic table could be rationalized using an extension of Lewis’ postulates. In 1922, N. Bohr updated his model of the atom by assuming that certain numbers of electrons (for example 2, 8, and 18) corresponded to stable “closed shells.” In 1926, Schrödinger established a wave mechanical description of the hydrogen atom which was subsequently extended to polyelectron atoms. Pauli was the first to realize that the complicated numbers of electrons in closed shells can be reduced to the simple rule of *one* per state, if the electron states are defined using four quantum numbers. For this purpose he introduced a new two-valued quantum number, identified by Goudsmit and Uhlenbeck as electron spin. The resulting Pauli Exclusion Principle states that no two electrons in a single atom can have the same four quantum numbers; if n, l, and m_l are the same, m_s must be different such that the electrons have opposite spins.

The idea of shared electron pairs introduced by Lewis provided an effective qualitative picture of covalent bonding and it still forms the basis of the universal notation for chemical communication, but it was Heitler and London who in 1927 developed the first successful quantum mechanical expression for this bonding model. Initially they provided a description of the bonding in molecular hydrogen, but it was subsequently adapted to more complex molecules and its widespread applications were articulated with great conviction by Linus Pauling. An alternative molecular orbital description of chemical bonding originated from Burrau’s description of the hydrogen molecule ion and this model was subsequently widely developed by Mulliken and Lennard-Jones. The electrons occupy molecular orbitals which are delocalized over the whole molecule and were filled according to the Aufbau Principle and assigned quantum numbers according to the Pauli Exclusion Principle. The orbitals are calculated in a self-consistent fashion in a manner analogous to those developed previously for atomic orbitals and are based on linear combination of the atomic orbitals of the individual atoms. The number of molecular orbitals equals the number of atomic orbitals in the atoms being combined to form the molecule. A molecular orbital describes the behavior of one electron in the electric field generated by the nuclei and some average distribution of the other electrons. This approximation proved to be more amenable to computer programming than the valence bond model and was widely developed and used in increasingly less approximate forms from 1960 to 1990.

In the early 1970s, a new electronic structure approach emerged from the physics community and was described as density functional theory (DFT). The total energy of a molecule was expressed as a functional of the total electron density. Hohenburg and Kohn proved the unique relationship between electron density and energy, and Kohn and Sham put forward a practical variational DFT approach. Although calculations in solid-state physics had been reported since the 1970s, DFT was not considered accurate enough for calculations in quantum chemistry until the 1990s, when the approximations used in the theory were refined to more accurately describe the exchange and correlation interactions. Computational costs for *ab initio* DFT calculations are relatively low when compared to the valence bond and molecular orbital methods. DFT thus began to approach the goals of computational thermochemistry to calculate the energetic properties of chemical processes to an accuracy of 1 kcal mol^{-1}. The widespread acceptance of these methodologies by the chemical community led to Kohn and Pople sharing the Nobel Prize in Chemistry in 1998.

When in 2004 Volumes 112 and 113 of Structure and Bonding were devoted to the "Principles and Applications of Density Functional Theory in Inorganic Chemistry" the editors N. Kaltsoyanis and J.E. McGardy noted "It is difficult to overestimate the impact that Density Functional Theory has had on computational quantum chemistry over the last two decades. Indeed, this period has seen it grow from little more than a theoretical curiosity to become a central tool in the computational chemist's armory." In these volumes they described recent applications in inorganic and biochemistry and addressed key issues in spectroscopy, mechanistic studies, and magnetism.

As a tribute to the continuing success story of DFT, the current two volumes mark the fundamental advances made in the first decade of the twenty-first century and the first volume provides an appropriate global view of its applications to chemical reactivity. It is apparent that the DFT methodology has acquired new and significant insights into modeling chemical phenomena which will have implications for at least the next decade and the chapters highlight some of the most exciting current directions of DFT. The first chapter reviews how Bose–Einstein condensation may be modeled by extension of DFT models, leading to a generalization of chemical bonding by employing the recently proposed advanced bondon—the quantum particle in the chemical bonding field. Then, the ideas central to information theory have been used to advance new concepts in molecular orbital theory, which leads to the prediction of the chemical interactions by means of "cascade" probability propagations through atomic orbital intermediates and these results are discussed in the second chapter. In the next chapter "conventional" high-priority classes of molecular systems such as "soft" metal clusters or weakly bonded molecules are analyzed and classified from polarizability and hyperpolarizability perspectives and this has been facilitated with the aid of new graph and pattern recognition techniques. It is followed by a chapter amenable to specialist and nonspecialist alike about different chemical reactivity descriptors such as electronegativity, chemical hardness, chemical softness, and Fukui functions. The combination of these concepts is conceptually responsible for helping chemists understand chemical stability and reactivity. The final chapter is dedicated to a historical view of the chemical indices and allied principles, with special references to hard and soft acids and bases (HSAB). This principle which was discussed at some length in early volumes of Structure and Bonding ultimately plays an important unifying role in qualitative descriptions of Lewis acid–base interactions and for rationalizing the rates of chemical reactions.

In summary, the volume critically reviews the applications of DFT to chemical reactivity from a structural and bonding perspective and aims to undermine those historical differences between physics and chemistry which were mentioned in the opening sentence.

We thank the contributors to this volume for the consistent efforts they have made in writing high-class scientific reviews and for providing the readers with a broad perspective which has revealed the widespread uses of DFT in interpreting chemical reactivity. MVP acknowledges the research and editing facilities provided for the present volume by the Romanian Education and Research Ministry within the project CNCS-UEFISCDI-TE-16/2010-2013. MVP and DMPM sincerely thank the Springer team and in particular Marion Hertel, Ursula Gramm, Elizabeth Hawkins, and Tanja Jaeger for professionally supervising the production of the Structure and Bonding series in general and of this volume in particular.

Timişoara, Romania Mihai V. Putz
Oxford, UK D. Michael P. Mingos

Contents

Struct Bond (2012) 149: 1–50
DOI: 10.1007/978-3-642-32753-7_1

Density Functional Theory of Bose–Einstein Condensation: Road to Chemical Bonding Quantum Condensate

Mihai V. Putz

Abstract The marriage of the celebrated fermionic and bosonic theories of matter that are Density Functional Theory and Bose–Einstein Condensation, respectively, is presented at the conceptual level of the Hohenberg–Kohn–Sham equations and energy functionals. Particular attention is given to the general formulation of the exchange-correlation within the context of perturbation theory at finite temperature, then specialized to the analytical exchange functional in terms of ordering parameter dependency by employing the Bogoliubov transformation (or the first-order main-field expansion); this further provides the appropriate framework within which the celebrated Gross–Pitaevsky BEC equation is recovered. This nonlinear generalization of the Schrödinger equation was then employed in first providing the DFT–BEC connection at the level of the Thomas–Fermi approximation; such connections were further used in generalizing the classical Heitler–London formalism and bonding–antibonding equations of homopolar chemical bonding with the aid of the mass quantification of the quantum particle of the chemical bonding field—the bondons. Actually, two branches of bosonic–bondonic condensate were identified as physical and chemical bonding BEC, both with bonding and antibonding features. The paradigmatic DFT–BEC application to the bonding interaction in hydrogen and helium molecular systems confirms both the physical and chemical forbidden bindings in He–He interactions, while allowing only the chemical interaction in the H–H system, though with bonding and antibonding quantum condensates below the "normal" ground-state potential of H_2. Further exciting perspectives at both physical and chemical levels, for conceptual and practical realizations of DFT–BEC systems, are in this review opened toward searching for unified fermionic–bosonic manifestations in nature.

M.V. Putz (✉)
Biology-Chemistry Department, West University of Timişoara, Pestalozzi Str. 16, 300115 Timişoara, Romania

Free University of Berlin, Theoretical Physics Institute, Arnimallee 14, 14195 Berlin, Germany
e-mail: mv_putz@yahoo.com; mvputz@cbg.uvt.ro

Keywords Bose–Einstein condensation · Density functional theory · Exchange functional · Gross–Pitaevsky equation · Thomas–Fermi approximation · Chemical action · Chemical hardness · Bondons · Chemical bond · Homopolar interaction · Hydrogen system · Helium system

Contents

1 Introduction

From the very inception of quantum mechanics, low-temperature physics in general and matter superfluidity and condensation phenomena in particular emerged and continuously achieved milestones in both theoretical and experimental studies. It its fascinating that the last 100 years of history have already generated such giant theoretical contributors as Bose [1], Einstein [2], London [3], Landau [4], Bogoliubov [5], Feynman [6], Penrose and Onsager [7], Lee, Huang and Yang [8], Pitaevsky [9], and Gross [10], paralleling the spectacular experimental breakthroughs achieved by no less ingenious laboratory leaders such as Kamerlingh-Onnes [11], Kapitza [12], Osheroff [13], Leggett [14], Cohen-Tannoudji [15], Wieman and Cornell [16], Ketterle [17], and Hänsch [18], whose work is now the forefront of the atomic and molecular physics worldwide. Initially regarded as a pathological exhibition of the noninteracting Bose gas, the Bose– Einstein condensation (BEC) is currently understood as a macroscopic occupation of a -single-particle state with a remarkable observable character. It firmly depends on the so-called critical temperature T_C; for instance, for homogeneous bosonic particles with mass m trapped in a 3D box, in the thermodynamic limit ($N \to \infty$, $V \to \infty$), the system fulfills the relationship [19]:

$$\rho \lambda_{\mathrm{dB}}^3 (T_{\mathrm{C}}) = \zeta(3/2) \cong 2.612, \tag{1}$$

thus linking it to the volume-density

$$\rho = \frac{N}{V}, \tag{2}$$

by means of the thermal de Broglie length:

$$\lambda_{\mathrm{dB}}(T) = \sqrt{\frac{2\pi\hbar^2}{mk_{\mathrm{B}}T}}, \tag{3}$$

and the Riemann zeta function:

$$\zeta(x) = \sum\nolimits_{k=1}^{\infty} k^{-x}. \tag{4}$$

The consequences are fundamental: on the one hand, one notes that the critical temperature may be either achieved through a higher density or a lower temperature; however, because within a high-density regime, systems may suffer chemical recombinations and atomic losses, while most chemical elements in high-density ground states are solid (or liquid as is the case of helium) rather than gases, dilute, low-temperature samples are more likely to achieve BEC. On the other hand, there is also evidence that the BEC phenomenology undoubtedly involves sample density.

It is at this point that the BEC calls for a density-based theory of matter; fortunately, such a theory exists under the celebrated Density Functional Theory (DFT), as developed by the eminent contributions of Kohn [20, 21], Parr [22], March [23], and Gross [24], among many others spanning both the quantum physical and chemical communities [25, 26]. Most notably, the DFT was originally designed to model electronic (strong) correlations in solid states and was even applied to superconductivity [26, 27]. However, superconductivity is considered to be related to BEC, a possibility recently revealed, for instance, by the observed BCS–BEC crossover [28–30] between the Bose–Einstein condensation and Barden–Cooper–Schrieffer theory of superconductivity [31], where the so-called Feshbach resonance [32] plays an important role. Therefore, the fundamental question arises as to whether DFT is only restricted to fermions or can be extended also to bosons. Thus, a DFT–BEC connection may be forged because both theories contain one-body wave functions $\psi(\boldsymbol{r})$ that are constructed in such a way that they rely on the same global density normalization condition, namely [19, 22],

$$N = \int \rho(\mathbf{r})\mathrm{d}\mathbf{r} = \int |\psi(\mathbf{r})|^2 \mathrm{d}\mathbf{r}. \tag{5}$$

Moreover, although DFT was formulated for Coulomb interparticle potentials, recent work reports that it may also accurately describe van der Waals interactions [33]; furthermore, it might be suitable for bosonic systems as well. It also admits many levels of comprehension [34], from homogeneous to local density or to the Thomas–Fermi approximation, which are also the current theoretical approaches of BEC [19]. Interestingly, only a very limited number of DFT–BEC studies have been reported [35–43], leaving much space to be filled by systematic research on a finite-temperature DFT approach to BEC. This work shows how such advancement is possible, in general and in detail, to model chemical bonding.

2 Bose–Einstein Condensates within Density Functional Theory: Ψ-DFT

2.1 *From Hohenberg–Kohn to Kohn–Sham Systems: Exchange-Correlation Functional*

From the beginning, one needs some prerequisites regarding the Density Functional Theory and its analytical connections to the statistical mechanics of Bosonic condensation. The main point is that in Bosonic condensates the energies and potential of the system are functionals of the couple $[\Psi(\mathbf{r}), \rho(\mathbf{r})]$ in terms of the *order parameter* $\Psi(\mathbf{r})$ and of the *superfluidic density* $\rho(\mathbf{r})$:

$$\begin{cases} \Psi(\mathbf{r}) \equiv \langle \psi(\mathbf{r}) \rangle_T \\ \rho(\mathbf{r}) \equiv \langle \psi^+(\mathbf{r}) \psi(\mathbf{r}) \rangle_T, \end{cases} \tag{6}$$

both of which are defined through the N-particles' ensemble statistical average at finite temperature ($T \neq 0$):

$$\begin{cases} \langle \bullet \rangle_T = \frac{1}{\mathcal{Z}} \mathrm{Tr}[\bullet e^{-\beta E[\psi]}] := \mathrm{Tr}[\rho_{\mathcal{Z}} \bullet] \\ \rho_{\mathcal{Z}} = \frac{e^{-\beta E[\psi]}}{\mathcal{Z}} = \frac{e^{-\beta E[\psi]}}{\mathrm{Tr}[e^{-\beta E[\psi]}]} \end{cases} \tag{7}$$

of the particle field $\psi(\mathbf{r})$, see Eq. (5), and $\beta = 1/k_B T$. Under these conditions, one can advance the main DFT–BEC idea according to which the real many-body system is represented by the so-called Hohenberg–Kohn grand-canonical potential (or free energy):

$$\Omega_{\mathrm{HK}}[\Psi, \rho] = F_{\mathrm{HK}}[\Psi, \rho] + \int d\mathbf{r}[V_{\mathrm{ext}}(\mathbf{r}) - \mu]\rho(\mathbf{r}) + D_{\mathrm{ext}}[\Psi, \rho], \tag{8}$$

with

$$\begin{cases} F_{\mathrm{HK}}[\Psi, \rho] = \left\langle T_{\mathrm{HK}} + W + \frac{\ln \rho_{\mathcal{Z}}}{\beta} \right\rangle_T \\ D_{\mathrm{ext}}[\Psi, \rho] = \int d\mathbf{r}[\Psi^+(\mathbf{r}) J(\mathbf{r}) + J^*(\mathbf{r}) \Psi(\mathbf{r})], \end{cases} \tag{9}$$

where F_{HK} is the Hohenberg–Kohn functional, $V_{ext}(\mathbf{r})$ is the external applied potential, μ is the chemical potential, W is the interparticle pair interaction, T_{HK} is the kinetic energy of the real system, and $J(\mathbf{r})$ is the external source (or current) that assures the observable character of the DFT–BEC approach. In the same manner, if one considers the Kohn–Sham (KS) system a noninteracting many-body system, it has by its very construction the same density and order parameter as the (HK) superfluid system,

$$\begin{cases} \rho(\mathbf{r}) \overset{!}{=} \rho_{KS}(\mathbf{r}) := \langle \psi^+(\mathbf{r})\psi(\mathbf{r})\rangle_T^{KS} \\ \Psi(\mathbf{r}) \overset{!}{=} \Psi_{KS}(\mathbf{r}) := \langle \psi(\mathbf{r})\rangle_T^{KS} \end{cases}, \tag{10}$$

though with the free energy specifically unfolding as follows:

$$\Omega_{KS}[\Psi,\rho] = F_{KS}[\Psi,\rho] + \int d\mathbf{r}[V_{KS}(\mathbf{r}) - \mu]\rho(\mathbf{r}) + D_{KS}[\Psi,\rho], \tag{11}$$

with the following components:

- The Kohn–Sham functional employing the actual KS-statistical average:

$$F_{KS}[\Psi,\rho] = \left\langle T_{KS} + \frac{\ln \rho_{Z}^{KS}}{\beta} \right\rangle_T^{KS} \tag{12}$$

- The Kohn–Sham potential:

$$V_{KS}(\mathbf{r}) = V_{ext}(\mathbf{r}) + V_{Hartree}(\mathbf{r}) + V_{XC}(\mathbf{r}), \tag{13}$$

with the introduced Hartree potential

$$V_{Hartree}(\mathbf{r}) = \int d\mathbf{r}'\rho(\mathbf{r}')W(\mathbf{r},\mathbf{r}'), \tag{14}$$

and the *exchange-correlation* potential contribution

$$V_{XC}(\mathbf{r}) = \frac{\delta E_{XC}[\Psi(\mathbf{r}),\Psi^+(\mathbf{r}),\rho(\mathbf{r})]}{\delta\rho}, \tag{15}$$

- The Kohn–Sham current potential:

$$D_{KS}[\Psi,\rho] = \int d\mathbf{r}\left[\Psi^+(\mathbf{r})J_{KS}(\mathbf{r}) + J_{KS}^*(\mathbf{r})\Psi(\mathbf{r})\right], \tag{16}$$

that is related to the external one of Eq. (9) through the exchange-correlation term:

$$D_{XC} = D_{KS} - D_{ext}, \tag{17}$$

while the KS current is defined analogously to the KS potential Eq. (15):

$$J_{KS}[\Psi,\rho](\mathbf{r}) = \frac{\delta E_{XC}[\Psi(\mathbf{r}),\Psi^{+}(\mathbf{r}),\rho(\mathbf{r})]}{\delta\Psi^{+}}. \tag{18}$$

Altogether, one may see that the KS grand-canonical potential/functional Eq. (11), with all of its components Eqs. (12)–(18), formally resembles the real KH functional Eq. (8) but for a different level of interaction. Therefore, the interaction is realized through their difference; it takes the form:

$$\Delta\Omega[\Psi,\rho] = \Omega_{HK}[\Psi,\rho] - \Omega_{KS}[\Psi,\rho],$$

$$= \int (V_{ext}(\mathbf{r}) - V_{KS}(\mathbf{r}))\rho(\mathbf{r})d\mathbf{r} - D_{XC}[\Psi,\rho] + E_{Hartree}[\Psi,\rho] + E_{XC}[\Psi,\rho], \tag{19}$$

providing the exchange-correlation functional within the DFT–BEC finite-temperature perturbation framework [35, 44]:

$$E_{XC}[\Psi,\rho] = -\frac{1}{\beta}\sum_{n=1}^{\infty}\left(-\frac{1}{\hbar}\right)^{n}\int_{0}^{\hbar}d\tau_{1}\ldots\int_{0}^{\hbar\beta}d\tau_{n}\left\langle \hat{T}\begin{bmatrix}(W-V_{Hartree}-V_{XC}-D_{XC})(\tau_{1}) \\ \ldots(W-V_{Hartree}-V_{XC}-D_{XC})(\tau_{n})\end{bmatrix}\right\rangle_{T}^{KS}$$
$$+\int (V_{Hartree}(\mathbf{r})+V_{XC}(\mathbf{r}))\rho(\mathbf{r})d\mathbf{r}+D_{XC}[\Psi,\rho]-E_{Hartree}[\Psi,\rho], \tag{20}$$

with

$$E_{Hartree}[\Psi,\rho] = \frac{1}{2}\int d\mathbf{r}d\mathbf{r}'\rho(\mathbf{r})W(\mathbf{r},\mathbf{r}')\rho(\mathbf{r}'). \tag{21}$$

Now, one may employ the exchange-correlation information to formulate the DFT–BEC equations, targeting the bosonic condensate density and the related density functional quantities [35, 44]:

$$\begin{cases}\left[-\frac{\hbar^{2}\nabla_{\mathbf{r}}^{2}}{2m} + V_{KS}(\mathbf{r}) - \mu\right]\varphi_{k}(\mathbf{r}) = E_{k}\varphi_{k}(\mathbf{r}) \\ \left[-\frac{\hbar^{2}\nabla_{\mathbf{r}}^{2}}{2m} + V_{KS}(\mathbf{r}) - \mu\right]\Psi(\mathbf{r}) = -J_{KS}(\mathbf{r}) \\ \rho(\mathbf{r}) = |\Psi(\mathbf{r})|^{2} + \sum_{k\neq 0}|\varphi_{k}(\mathbf{r})|^{2}\frac{1}{\exp(\beta E_{k})-1}.\end{cases} \tag{22}$$

The solutions to this system Eq. (22) with Eqs. (15), (18), and (20) will give the condensate density, i.e., the ground-state density, and the excited spectra for the thermal fluctuations in a single "shot," thus revealing the advantage of DFT in BEC reformulation. However, due to the practically insurmountable difficulty in solving this self-consistent system for general interparticle and external trapping potentials, one should further rely on custom DFT approximation techniques specialized for BEC.

2.2 From Exchange Correlation to Exchange Density Functionals of Bose–Einstein Condensate

Because the current aim is to address dilute gas states, i.e., the first-order approximation of DFT–BEC under the first-order parameter expansion, the g-DFT form with

$$g = |\Psi(\mathbf{r})|^2 \equiv \langle\psi(\mathbf{r})\rangle_T^2, \tag{23}$$

is wholly sufficient, which is also the case for the exchange-correlation energy above, Eq. (20), such that only the *exchange* energy will be addressed. Therefore, one can express the exchange BEC-density functional explicitly as [35, 44]:

$$\begin{aligned}
E_X[\Psi,\rho] &= \frac{1}{\beta\hbar}\int_0^{\hbar\beta} d\tau\left\langle \hat{\mathrm{T}}[(W - V_{\mathrm{Hartree}} - V_{\mathrm{XC}} - D_{\mathrm{XC}})(\tau)]\right\rangle_T^{\mathrm{KS}} \\
&\quad + \int (V_{\mathrm{Hartree}}(\mathbf{r}) + V_{\mathrm{XC}}(\mathbf{r}))\rho(\mathbf{r})d\mathbf{r} + \int D_{\mathrm{XC}}[\Psi,\rho]d\mathbf{r} - E_{\mathrm{Hartree}}[\Psi,\rho] \\
&= \frac{1}{\beta\hbar}\int_0^{\hbar\beta} d\tau\left\langle \hat{\mathrm{T}}[W(\tau)]\right\rangle_T^{\mathrm{KS}} - E_{\mathrm{Hartree}}[\Psi,\rho] \\
&\quad \underbrace{-\frac{1}{\beta\hbar}\int_0^{\hbar\beta} d\tau\left\langle \hat{\mathrm{T}}[V_{\mathrm{Hartree}}(\tau)]\right\rangle_T^{\mathrm{KS}}}_{\int V_{\mathrm{Hartree}}(\mathbf{r})\rho(\mathbf{r})d\mathbf{r}} \underbrace{-\frac{1}{\beta\hbar}\int_0^{\hbar\beta} d\tau\left\langle \hat{\mathrm{T}}[V_{\mathrm{XC}}(\tau)]\right\rangle_T^{\mathrm{KS}}}_{\int V_{\mathrm{XC}}(\mathbf{r})\rho(\mathbf{r})d\mathbf{r}} \\
&\quad \underbrace{-\frac{1}{\beta\hbar}\int_0^{\hbar\beta} d\tau\left\langle \hat{\mathrm{T}}[D_{\mathrm{XC}}(\tau)]\right\rangle_T^{\mathrm{KS}}}_{\int D_{\mathrm{XC}}(\mathbf{r})\rho(\mathbf{r})d\mathbf{r}} + \int V_{\mathrm{Hartree}}(\mathbf{r})\rho(\mathbf{r})d\mathbf{r} \\
&\quad + \int V_{\mathrm{XC}}(\mathbf{r})\rho(\mathbf{r})d\mathbf{r} + \int D_{\mathrm{XC}}[\Psi,\rho]d\mathbf{r} \\
&= \frac{1}{\beta\hbar}\int_0^{\hbar\beta} d\tau\left\langle \hat{\mathrm{T}}[W(\tau)]\right\rangle_T^{\mathrm{KS}} - E_{\mathrm{Hartree}}[\Psi,\rho],
\end{aligned} \tag{24}$$

or in its practical form:

$$\begin{aligned}E_X[\Psi,\rho] &= \frac{1}{2}\int W(\mathbf{r},\mathbf{r}')\rho(\mathbf{r})\rho(\mathbf{r}')\mathrm{d}\mathbf{r}\mathrm{d}\mathbf{r}' - E_{\text{Hartree}}[\Psi,\rho]\\ &= \frac{1}{2}\int W(\mathbf{r},\mathbf{r}')\langle\psi^+(\mathbf{r})\psi^+(\mathbf{r}')\psi(\mathbf{r})\psi(\mathbf{r}')\rangle_T^{\mathrm{KS}}\, d\mathbf{r}\, d\mathbf{r}' - E_{\text{Hartree}}[\Psi,\rho],\end{aligned} \tag{25}$$

while retaining the main Ψ-DFT factors, particularly the superfluidic density Eq. (6).

Next, one may evaluate the average interparticle interaction in Eq. (25) by unfolding it through the Bogoliubov transformations:

$$\psi(\mathbf{r}) = \Psi(\mathbf{r}) + \tilde{\psi}(\mathbf{r}), \tag{26}$$

$$\psi^+(\mathbf{r}) = \Psi^+(\mathbf{r}) + \tilde{\psi}^+(\mathbf{r}), \tag{27}$$

which immediately yields

$$\begin{aligned}&\psi^+(\mathbf{r})\psi^+(\mathbf{r}')\psi(r)\psi(\mathbf{r}')\\ &= |\Psi(\mathbf{r})|^2|\Psi(\mathbf{r}')|^2 + \tilde{\psi}^+(\mathbf{r})\tilde{\psi}^+(\mathbf{r}')\tilde{\psi}(\mathbf{r}')\tilde{\psi}(\mathbf{r}')\\ &\quad + 2|\Psi(\mathbf{r})|^2\tilde{\psi}^+(\mathbf{r}')\tilde{\psi}(\mathbf{r}') + 2\Psi(\mathbf{r})\Psi^+(\mathbf{r}')\tilde{\psi}^+(\mathbf{r})\tilde{\psi}(\mathbf{r}')\\ &\quad + \Psi(\mathbf{r})\Psi(\mathbf{r}')\tilde{\psi}^+(\mathbf{r})\tilde{\psi}^+(\mathbf{r}') + \Psi^+(\mathbf{r})\Psi^+(\mathbf{r}')\tilde{\psi}(\mathbf{r})\tilde{\psi}(\mathbf{r}')\\ &\quad + 2|\Psi(\mathbf{r})|^2\Psi(\mathbf{r}')\underbrace{\tilde{\psi}^+(\mathbf{r}')}_{!} + 2|\Psi(\mathbf{r})|^2\Psi^+(\mathbf{r}')\underbrace{\tilde{\psi}(\mathbf{r}')}_{!}\\ &\quad + 2\Psi^+(\mathbf{r})\underbrace{\tilde{\psi}^+(\mathbf{r}')\tilde{\psi}(\mathbf{r}')\tilde{\psi}(\mathbf{r})}_{!} + 2\Psi(\mathbf{r})\underbrace{\tilde{\psi}(\mathbf{r})\tilde{\psi}^+(\mathbf{r}')\tilde{\psi}(\mathbf{r}')}_{!}.\end{aligned} \tag{28}$$

When the grand statistical average is applied to expression (28), one obtains that [44]

- The order-parameter-containing term behaves like a scalar and is left untouched:

$$\left\langle|\Psi(\mathbf{r})|^2|\Psi(\mathbf{r}')|^2\right\rangle_T^{\mathrm{KS}} = |\Psi(\mathbf{r})|^2|\Psi(\mathbf{r}')|^2. \tag{29}$$

- The terms with the conjugated product of the fluctuation (or thermal) field produces the associate density:

$$\left\langle 2|\Psi(\mathbf{r})|^2\tilde{\psi}^+(\mathbf{r}')\tilde{\psi}(\mathbf{r}')\right\rangle_T^{\mathrm{KS}} = 2|\Psi(\mathbf{r})|^2\tilde{\rho}(\mathbf{r}'). \tag{30}$$

- At different spatial points, the conjugate products of the fluctuating fields produce the Green function (at the same time/temperature), according to its basic definition:

$$G(\mathbf{r}\tau;\mathbf{r}'\tau) = -\left\langle \tilde{\psi}^{+}(\mathbf{r})\tilde{\psi}(\mathbf{r}')\right\rangle_{T}^{\mathrm{KS}}, \tag{31}$$

such that one has, for the two specific terms in Eq. (28),

$$\left\langle 2\Psi(\mathbf{r})\Psi^{+}(\mathbf{r}')\tilde{\psi}^{+}(\mathbf{r})\tilde{\psi}(\mathbf{r}')\right\rangle_{T}^{\mathrm{KS}} = -2\Psi(\mathbf{r})\Psi^{+}(\mathbf{r}')G(\mathbf{r}\tau;\mathbf{r}'\tau), \tag{32}$$

$$\left\langle \tilde{\psi}^{+}(\mathbf{r})\tilde{\psi}^{+}(\mathbf{r}')\tilde{\psi}(\mathbf{r}')\tilde{\psi}(\mathbf{r})\right\rangle_{T}^{\mathrm{KS}} = G(\mathbf{r}\tau;\mathbf{r}'\tau)G(\mathbf{r}'\tau;\mathbf{r}\tau). \tag{33}$$

- The sum of the conjugated terms that cannot be obtained by averaging (trace operation) either the Green function or the thermal density is considered to generate a double real part of the associated product of thermal/fluctuation densities, which by averaging is divided by a factor of 2 and, in its simple form, produces

$$\left\langle \Psi(\mathbf{r})\Psi(\mathbf{r}')\tilde{\psi}^{+}(\mathbf{r})\tilde{\psi}^{+}(\mathbf{r}') + \Psi^{+}(\mathbf{r})\Psi^{+}(\mathbf{r}')\tilde{\psi}(\mathbf{r})\tilde{\psi}(\mathbf{r}')\right\rangle_{T}^{\mathrm{KS}} = \tilde{\rho}(\mathbf{r})\tilde{\rho}(\mathbf{r}'). \tag{34}$$

- The terms containing odd powers of fluctuating/thermal field in Eq. (28) average to zero because by construction they cause the statistical average of the fluctuating/thermal field to tend to zero,

$$\left\langle \tilde{\psi}(\mathbf{r})\right\rangle_{T} = 0 = \left\langle \tilde{\psi}^{+}(\mathbf{r})\right\rangle_{T}, \tag{35}$$

i.e., this does not produce any Bose–Einstein condensate that is characterized by the order parameter Eq. (6); thus, one has

$$\begin{aligned} 0 &= \left\langle 2|\Psi(\mathbf{r})|^{2}\Psi(\mathbf{r}')\tilde{\psi}^{+}(\mathbf{r}')\right\rangle_{T}^{\mathrm{KS}} = \left\langle 2|\Psi(\mathbf{r})|^{2}\Psi^{+}(\mathbf{r}')\tilde{\psi}(\mathbf{r}')\right\rangle_{T}^{\mathrm{KS}} \\ &= \left\langle 2\Psi^{+}(\mathbf{r})\tilde{\psi}^{+}(\mathbf{r}')\tilde{\psi}(\mathbf{r}')\tilde{\psi}(\mathbf{r})\right\rangle_{T}^{\mathrm{KS}} = \left\langle 2\Psi(\mathbf{r})\tilde{\psi}(\mathbf{r})\tilde{\psi}^{+}(\mathbf{r}')\tilde{\psi}(\mathbf{r}')\right\rangle_{T}^{\mathrm{KS}}. \end{aligned} \tag{36}$$

In summary, with Eqs. (28)–(36), the exchange expression (25) is now specifically

$$\begin{aligned} E_{X}[\Psi,\rho] = \frac{1}{2}\int d\mathbf{r}d\mathbf{r}'W(\mathbf{r},\mathbf{r}')\Big\{&|\Psi(\mathbf{r})|^{2}|\Psi(\mathbf{r}')|^{2} + 2|\Psi(\mathbf{r})|^{2}\tilde{\rho}(\mathbf{r}') + \tilde{\rho}(\mathbf{r})\tilde{\rho}(\mathbf{r}') \\ &- 2\Psi(\mathbf{r})\Psi^{+}(\mathbf{r}')G(\mathbf{r}\tau;\mathbf{r}'\tau) + G(\mathbf{r}\tau;\mathbf{r}'\tau)G(\mathbf{r}'\tau;\mathbf{r}\tau)\Big\} - E_{\mathrm{Hartree}}[\Psi,\rho]. \end{aligned} \tag{37}$$

Considering the relationships between the superfluid, condensate, and fluctuating densities in the *main-field approximation of the first order*, one has, from Eq. (26), for instance,

$$\rho(\mathbf{r}) = |\Psi(\mathbf{r})|^2 + \tilde{\rho}(\mathbf{r}), \tag{38}$$

so that recognizing the vanishing of the first order for the fluctuation/thermal field

$$\tilde{\psi}(\mathbf{r}) \to 0. \tag{39}$$

With condition Eq. (38), the first three integrand terms of Eq. (37) become

$$\begin{aligned}
&|\Psi(\mathbf{r})|^2|\Psi(\mathbf{r}')|^2 + 2|\Psi(\mathbf{r})|^2\tilde{\rho}(\mathbf{r}') + \tilde{\rho}(\mathbf{r})\tilde{\rho}(\mathbf{r}')\\
&= |\Psi(\mathbf{r})|^2|\Psi(\mathbf{r}')|^2 + 2|\Psi(\mathbf{r})|^2\left(\rho(\mathbf{r}') - |\Psi(\mathbf{r}')|^2\right)\\
&\quad + \left(\rho(\mathbf{r}) - |\Psi(r)|^2\right)\left(\rho(\mathbf{r}') - |\Psi(\mathbf{r}')|^2\right)\\
&= \rho(\mathbf{r})\rho(\mathbf{r}').
\end{aligned} \tag{40}$$

Therefore providing the Hartree integral Eq. (21) in Eq. (37) leads to the final result for the exchange functional:

$$E_X[\Psi,\rho] = \frac{1}{2}\int d\mathbf{r}d\mathbf{r}'W(\mathbf{r},\mathbf{r}')\left\{-2\Psi(\mathbf{r})\Psi^+(\mathbf{r}')G(\mathbf{r}\tau;\mathbf{r}'\tau) + |G(\mathbf{r}\tau;\mathbf{r}'\tau)|^2\right\}. \tag{41}$$

Note that if one would like to visualize Eq. (41) in Feynman-like diagrams, the expression is worth being rewritten first as a Green function series expansion [35, 44]:

$$E_X[\Psi,\rho] = \int d\mathbf{r}d\mathbf{r}'W(\mathbf{r},\mathbf{r}')\left\{\Psi(\mathbf{r})\Psi^+(\mathbf{r}')g(\mathbf{r}\tau;\mathbf{r}'\tau) + \frac{1}{2}|g(\mathbf{r}\tau;\mathbf{r}'\tau)|^2\right\}, \tag{42}$$

where we consider the Green-function re-notation

$$G(\mathbf{r}\tau;\mathbf{r}'\tau) = -g(\mathbf{r}\tau;\mathbf{r}'\tau), \tag{43}$$

such that the final Feynman diagrammatical result is

$E_X[\Psi,\rho] =$ 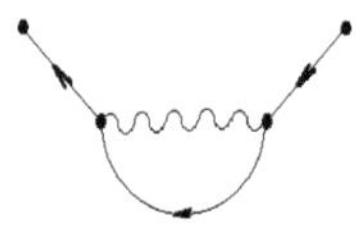 $+\frac{1}{2}$ 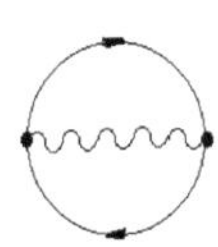 (44)

One may find this result instructive when going to higher orders based on systematical Feynman diagram expansion. For instance, in the second-order expansion, one would also expect a correlation contribution, although one could claim that such contributions are of no use for bosonic systems unless they are very inhomogeneous.

2.3 Recovering Gross–Pitaevsky Equation Within Local Density Approximation of Ψ-DFT

At this point, one would like to verify whether the Ψ-DFT in general and the Kohn–Sham equation in particular provide a point of contact with the specific Bose–Einstein condensation equation, in the mean-field approximation, as the Gross–Pitaevsky does. To this end, one may start by noticing the difference between *perturbation* and *interaction* when speaking about the two DFT systems [43, 44]:

- The natural or Hohenberg–Kohn system with external $V_{ext}(\mathbf{r})$ and interacting $W(\mathbf{r},\mathbf{r}')$ potentials, the last one being included in the HK-functional F_{HK}; the specific auxiliary order-parameter-related potential D_{HK} is considered formally because ultimately it must be set to zero because its coupling currents are vanishing ($J_{HK} \to 0$) from a macroscopic point of view (the equation of motion). From this perspective, the grand potential, see Eq. (8), is viewed in a more convenient way by unifying the HK notations for the external, i.e., "*ext.* = HK"
- The noninteracting or Kohn–Sham (auxiliary) construction with artificial KS potential V_{KS}, which has an effective potential role. This also includes the external or HK potential but with no longer any interaction potential, though with the same density (and order parameter) as that characterizing the HK system; in this case, which addresses a noninteracting state, for the order parameter to exist, the associated potential D_{KS} no longer vanishes; it is characterized by the currents J_{KS}, which play an important role in the KS equation itself, as will soon be made evident. For convenience, also for KS, one will unify all of the notations in the grand canonical potential as in Eq. (11).

In particular, the grand canonical potential Eq. (11) helps in producing the Ψ-KS equation by applying the first equation of the stationary system:

$$\begin{cases} \dfrac{\delta}{\delta \Psi^{+}} \Omega_{KS} = 0 \\ \dfrac{\delta}{\delta \rho} \Omega_{KS} = 0. \end{cases} \tag{45}$$

This means that one should first rewrite Eq. (11) in terms of the order parameter with the help of the Bogoliubov transformation Eqs. (26) and (27), while taking into account the vanishing rule for the integrals containing odd powers of the fluctuating (or thermal) field. Therefore, in the zero-temperature limit $T \to 0; \beta \to \infty$, the grand canonical potential Eq. (11) takes the following form [35, 43, 44]:

$$\Omega_{KS}[\Psi,\tilde{\psi}] = \int d\mathbf{r}\Psi^{+}(\mathbf{r})\hat{T}\Psi(\mathbf{r}) + \int d\mathbf{r}\tilde{\psi}^{+}(\mathbf{r})\hat{T}\tilde{\psi}(\mathbf{r})$$
$$+ \int d\mathbf{r}\Psi^{+}(\mathbf{r})[V_{KS}(\mathbf{r}) - \mu]\Psi(\mathbf{r}) + \int d\mathbf{r}\tilde{\psi}^{+}(\mathbf{r})[V_{KS}(\mathbf{r}) - \mu]\tilde{\psi}(\mathbf{r})$$
$$+ \int d\mathbf{r}\left[\Psi^{+}(\mathbf{r})J_{KS}(\mathbf{r}) + J^{*}_{KS}(\mathbf{r})\Psi(\mathbf{r})\right], \tag{46}$$

and its functional derivative with respect to the order parameter, the first equation of (45), transforms the Ψ-KS equation into the following:

$$\left(-\frac{\hbar^{2}\nabla^{2}}{2m} + V_{KS}(\mathbf{r}) - \mu\right)\Psi(\mathbf{r}) = -J_{KS}(\mathbf{r}), \tag{47}$$

making the expression ready to be employed in a real system characterized by the Hohenberg–Kohn grand potential Eq. (8). However, before proceeding further, one should note the second equation of the system Eq. (45) will produce the Schrödinger equation associated with the thermal part $(\tilde{\psi})$ of the (superfluid) system, as previously proved, see [35, 44]. Instead, new physical information is expected from the order parameter equation, i.e., the Ψ-KS Eq. (47).

Returning to the natural or HK system, its great potential Eq. (8) must fulfill the same optimization conditions as the KS system, namely,

$$\begin{cases} \dfrac{\delta}{\delta\Psi^{+}}\Omega_{HK} = 0 \\ \dfrac{\delta}{\delta\rho}\Omega_{HK} = 0. \end{cases} \tag{48}$$

Now, to make use of the KS variational equations (45), one must employ (8) as follows [35, 43, 44]:

$$\Omega_{HK}[\Psi,\rho] = \Omega_{KS}[\Psi,\rho] + \int f_{int}(\Psi,\rho)d\mathbf{r}$$
$$- \int d\mathbf{r}[V_{KS}(\mathbf{r}) - V_{HK}(\mathbf{r})]\rho(\mathbf{r}) - \int d\mathbf{r}\left[\Psi^{+}(\mathbf{r})J_{KS}(\mathbf{r}) + J^{*}_{KS}(\mathbf{r})\Psi(\mathbf{r})\right], \tag{49}$$

where the *local density approximation* (LDA) for the interaction term is introduced, namely,

$$F_{int}[\Psi,\rho] = \int f_{int}(\Psi,\rho)d\mathbf{r}, \tag{50}$$

while recalling that here the interparticle energy W stands for the identified interaction; the way to incorporate its double coordinate dependency into the LDA framework will be soon revealed. For the moment, let us apply Eqs. (48) to (49) while counting for Eq. (45) as well; then, we will have, respectively,

$$\begin{cases} \frac{\delta}{\delta\Psi^{+}} f_{\text{int}}(\Psi,\rho) = J_{\text{KS}} \\ \frac{\delta}{\delta\rho} f_{\text{int}}(\Psi,\rho) = V_{\text{KS}}(\mathbf{r}) - V_{\text{HK}}(\mathbf{r}). \end{cases} \tag{51}$$

Now, one has to substitute the KS current and potential from Eq. (51) into Eq. (47) to obtain the working form of Ψ-KS:

$$\left(-\frac{\hbar^2\nabla^2}{2m} + V_{\text{ext}}(\mathbf{r}) - \mu + \frac{\delta}{\delta\rho} f_{\text{int}}(\Psi,\rho)\right)\Psi(\mathbf{r}) = -\frac{\delta}{\delta\Psi^{+}} f_{\text{int}}(\Psi,\rho), \tag{52}$$

where, eventually, we reapply the "HK = ext." notation for precision. Equation (52) definitely holds for a modified Schrödinger equation, being specific to condensates (order parameter) or to superfluids in its mean-field approximation.

The next task is to further adapt Eq. (52) to the present case of interaction. One can identify, based on the previous exchange relation (37), the actual interaction functional [35, 43, 44]:

$$\begin{aligned} &F_{\text{int}}[\Psi,\rho] \\ &= \frac{1}{2}\int d\mathbf{r}d\mathbf{r}' W(\mathbf{r},\mathbf{r}') \left\{ \begin{array}{l} |\Psi(\mathbf{r})|^2|\Psi(\mathbf{r}')|^2 + 2|\Psi(\mathbf{r})|^2\tilde{\rho}(\mathbf{r}') + \tilde{\rho}(\mathbf{r})\tilde{\rho}(\mathbf{r}') \\ -2\Psi(\mathbf{r})\Psi^{+}(\mathbf{r}')G(\mathbf{r}\tau;\mathbf{r}'\tau) + G(\mathbf{r}\tau;\mathbf{r}'\tau)G(\mathbf{r}'\tau;\mathbf{r}\tau) \end{array} \right\}. \end{aligned} \tag{53}$$

Equation (53) becomes of the LDA type Eq. (50) in the case of bilocal-to-local transformation of the interaction according to the delta-function dependency of the order parameter:

$$W(\mathbf{r},\mathbf{r}') \to g\delta(\mathbf{r}-\mathbf{r}'), \tag{54}$$

in which case its coupling with the Green function is as follows:

$$\int d\mathbf{r}'\delta(\mathbf{r}-\mathbf{r}')G(\mathbf{r}\tau;\mathbf{r}'\tau) = -\int d\mathbf{r}'\delta(\mathbf{r}-\mathbf{r}')\left\langle \tilde{\psi}^{+}(\mathbf{r})\tilde{\psi}(\mathbf{r}')\right\rangle_T^{\text{KS}} = -\langle\tilde{\rho}(\mathbf{r})\rangle_T^{\text{KS}}, \tag{55}$$

and the interaction functional becomes

$$F_{\text{int}}[\Psi,\rho] = \frac{g}{2}\int d\mathbf{r} \left\{ \begin{array}{l} \Psi^{+2}(\mathbf{r})\Psi^2(\mathbf{r}) + 2|\Psi(\mathbf{r})|^2\tilde{\rho}(\mathbf{r}) + \tilde{\rho}^2(\mathbf{r}) \\ +2\Psi(\mathbf{r})\Psi^{+}(\mathbf{r})\langle\tilde{\rho}(\mathbf{r})\rangle_T^{\text{KS}} + \langle\tilde{\rho}(\mathbf{r})\rangle_T^{\text{KS}}\langle\tilde{\rho}(\mathbf{r})\rangle_T^{\text{KS}} \end{array} \right\}. \tag{56}$$

Next, because we would like to arrive at an order parameter and density dependence, we substitute the fluctuating density from Eq. (38) into Eq. (56) to obtain [35, 44]:

$$
\begin{aligned}
F_{\text{int}}[\Psi,\rho] &= \frac{g}{2}\int \mathrm{d}\mathbf{r}\left\{\begin{array}{l} \rho^2(\mathbf{r})+2\Psi(\mathbf{r})\Psi^+(\mathbf{r})\langle[\rho(\mathbf{r})-|\Psi(\mathbf{r})|^2]\rangle_T^{\text{KS}} \\ +\langle[\rho(\mathbf{r})-|\Psi(\mathbf{r})|^2]\rangle_T^{\text{KS}}\langle[\rho(\mathbf{r})-|\Psi(\mathbf{r})|^2]\rangle_T^{\text{KS}} \end{array}\right\} \\
&= \frac{g}{2}\int \mathrm{d}r\left\{\begin{array}{l} \rho(\mathbf{r})^2+2\Psi(\mathbf{r})\Psi^+(\mathbf{r})\langle\rho(\mathbf{r})\rangle_T^{\text{KS}}-2\Psi(\mathbf{r})\Psi^+(\mathbf{r})\langle|\Psi(\mathbf{r})|^2\rangle_T^{\text{KS}} \\ +\langle\rho(\mathbf{r})\rangle_T^{\text{KS}}\langle\rho(\mathbf{r})\rangle_T^{\text{KS}}-2\langle\rho(\mathbf{r})\rangle_T^{\text{KS}}\langle|\Psi(\mathbf{r})|^2\rangle_T^{\text{KS}}+\langle|\Psi(\mathbf{r})|^2\rangle_T^{\text{KS}}\langle|\Psi(\mathbf{r})|^2\rangle_T^{\text{KS}} \end{array}\right\}.
\end{aligned} \tag{57}
$$

At this point, the terms in Eq. (57) are employed as follows [43, 44]:

- The thermodynamic average over the order parameter may leave it unaffected because it characterizes the condensate order anyway; this is the case with the third term of Eq. (57):

$$
\left\langle|\Psi(\mathbf{r})|^2\right\rangle_T^{\text{KS}} = |\Psi(\mathbf{r})|^2. \tag{58}
$$

- The product of the two thermodynamic averages over the superfluid density reduces to that of the order parameter in the mean-field approximation:

$$
\langle\rho(\mathbf{r})\rangle_T^{\text{KS}}\langle\rho(\mathbf{r})\rangle_T^{\text{KS}} \cong \left\langle|\Psi(\mathbf{r})|^2\right\rangle_T^{\text{KS}}\left\langle|\Psi(\mathbf{r})|^2\right\rangle_T^{\text{KS}} = [\Psi^+(\mathbf{r})\Psi(\mathbf{r})]^2. \tag{59}
$$

- In the same framework of the mean-field approximation, for the last term of Eq. (57), the fourth power of the mean field is considered to produce the effect of the squared superfluid density, respectively,

$$
|\Psi(\mathbf{r})|^4 \cong \rho^2(\mathbf{r}). \tag{60}
$$

This way, expression (57) simplifies to

$$
F_{\text{int}}[\Psi,\rho] = \frac{g}{2}\int \mathrm{d}\mathbf{r}\left\{2\rho^2(\mathbf{r}) - [\Psi^+(\mathbf{r})\Psi(\mathbf{r})]^2\right\}. \tag{61}
$$

Through comparison with Eq. (50), the interaction density is then extracted using the expression:

$$
f_{\text{int}}(\Psi,\rho) = g\left\{\rho^2(\mathbf{r}) - \frac{1}{2}[\Psi^+(\mathbf{r})\Psi(\mathbf{r})]^2\right\}. \tag{62}
$$

Immediately, the derivatives needed in Eq. (51) are formed:

$$
\begin{cases} \dfrac{\delta}{\delta\rho}f_{\text{int}}(\Psi,\rho) = 2g\rho(\mathbf{r}) \\ \dfrac{\delta}{\delta\Psi^+}f_{\text{int}}(\Psi,\rho) = -g[\Psi^+(\mathbf{r})\Psi(\mathbf{r})]\Psi(\mathbf{r}), \end{cases} \tag{63}
$$

such that when they are substituted into Ψ-KS Eq. (51), they produce the mean-field equation:

$$\left(-\frac{\hbar^2\nabla^2}{2m}+V_{\mathrm{ext}}(\mathbf{r})-\mu+g[2\rho(\mathbf{r})-\Psi^{+}(\mathbf{r})\Psi(\mathbf{r})]\right)\Psi(\mathbf{r})=0. \tag{64}$$

Finally, one may once again apply the mean-field expansion (38) to evaluate the g-coupling in Eq. (64):

$$2\rho(\mathbf{r})-\Psi^{+}(\mathbf{r})\Psi(\mathbf{r})=\Psi^{+}(\mathbf{r})\Psi(\mathbf{r})+2\tilde{\psi}^{+}(\mathbf{r})\tilde{\psi}(\mathbf{r})+2\left[\Psi^{+}(\mathbf{r})\tilde{\psi}(\mathbf{r})+\Psi(\mathbf{r})\tilde{\psi}^{+}(\mathbf{r})\right]. \tag{65}$$

The last term of Eq. (65) vanishes through the left multiplication of Eq. (64) with $\Psi^{+}(\mathbf{r})$ followed by integration due to the appearance of the first-order fluctuating/thermal field, thus leaving Eq. (64) with the so-called modified Gross–Pitaevsky (or Hartree–Fock–Bogoliubov) equation [43, 44]:

$$\left(-\frac{\hbar^2\nabla^2}{2m}+V_{\mathrm{ext}}(\mathbf{r})-\mu+2g\tilde{\rho}(\mathbf{r})+g\Psi^{+}(\mathbf{r})\Psi(\mathbf{r})\right)\Psi(\mathbf{r})=0. \tag{66}$$

Equation (66) formally differs from the Landau version of refs. [4, 45] by the presence of the thermal density, $\tilde{\rho}(\mathbf{r})$. However, noting that the original Gross–Pitaevsky equation characterizes the ground state of the condensate field/density,

$$\left(-\frac{\hbar^2\nabla^2}{2m}+V_{\mathrm{ext}}(\mathbf{r})+g|\Psi(\mathbf{r})|^2\right)\psi(\mathbf{r})=\mu\Psi(\mathbf{r}), \tag{67}$$

it is viewed as the nonlinear (or generalized) Schrödinger equation for superfluid systems. Nevertheless, the main point here is that the Density Functional Theory in general and its order parameter version Ψ-KS fully agree with the conventional treatment of the Bose–Einstein condensate, a notion that is further generalized by recovering the Gross–Pitaevsky equation fashioned in the local density approximation when the interparticle potential is reduced to a point interaction, i.e., for a system with the interaction Hamiltonian:

$$\hat{H}_{\mathrm{int}}=\frac{g}{2}\int d\mathbf{r}\psi^{+}(\mathbf{r})\psi^{+}(\mathbf{r})\psi(\mathbf{r})\psi(\mathbf{r}). \tag{68}$$

Indeed, the local density approximation is the next topic discussed in this review.

3 Practical DFT–BEC Connections Within Thomas–Fermi Approximation

The practical implementation of BEC and of its mean-field approximation usually makes use of thermodynamic limit constraint, namely for systems with many-to-infinite number of particles ($N \to \infty$). The so-called Thomas–Fermi approximation [46, 47] may be used such that the potential and the interaction energies are larger than the kinetic energy, which can be therefore neglected in the stationary Gross–Pitaevsky equation, reducing it to the algebraic form [41, 42]:

$$|\psi|^2 \cong \frac{1}{g}[\mu - V(\mathbf{r})]. \tag{69}$$

Now, the discussion is whether this approximation is suitable for:

- BEC with a finite number of boson interactions, as is the practical case
- Single bosonic (bondonic) chemical bonding

The answer is fortunately given by a series of pioneering papers in physics philosophically questioning the significant difference in systems with more or infinitely more components [48–50]; the conclusion is simply that

- "more is different" [48]
  whereas
- "infinitely more is the same" [49]

These may be seen as nothing more than a re-definition of Fermi and Bose statistics, respectively. In other words, when more than one but a finite number of particles are present in a system, the statistical behavior is different than that of single-particle systems, while when an infinite number of particles comes into play, they all tend to behave as in a (the) single-particle case. Moreover, the Anderson assertion may be translated for chemical systems as

- "few electrons is different"
  with respect to their arrangement in atomic and molecular orbitals, while when they are regarded as bonding electrons in bondons, the issue becomes
- "pairing electrons is the same"
  which treats the formed bondons like single-bosonic systems, formally mapped into the bosonic condensation of an infinite number of bondons that behave the same as one!

Returning to the Thomas–Fermi approximation, the approximation is indeed in agreement with the thermodynamic limit for bosonic condensation and can also be used for chemical systems, for chemical bonds are present down to the limiting case of single-chemical-bond systems (as in the paradigmatic homopolar bindings of molecular hydrogen and helium, H_2 and He_2).

However, the finite N-dependency maybe restored from the Thomas–Fermi limit because the main equation of the Density Functional Theory holds [51, 52]:

$$\delta\left(E - \mu \int d\mathbf{r}\rho(\mathbf{r})\right) = 0. \tag{70}$$

This minimizes the total energy density functional under the constraint of conservation of the total number of electrons in the system, see Eq. (5), making Eq. (69) the striking *first connection between the custom* BEC *and* DFT formalisms of matter (we will specify later what quantities are associated with fermions and bosons in the mixed bosonic–fermionic systems of chemical bonding). However, even more, after expanding the total energy functional according to the Hohenberg–Kohn–Sham rule [20, 21]:

$$\begin{aligned} E[\rho] &= \int d\mathbf{r}V(\mathbf{r})\rho(\mathbf{r}) + E_{\text{kin}}[\rho] + E_{Coulomb}[\rho] + E_{\text{XC}}[\rho] \\ &= C_A[\rho] + F_{\text{HK}}[\rho], \end{aligned} \tag{71}$$

in terms of the Hohenberg–Kohn functional $F_{\text{HK}}[\rho]$ and of the chemical action $C_A[\rho]$ [41–44, 53], the variational DFT principle is released from the working relationship between the chemical potential, external potential, and the density functional derivative of kinetic, Coulombic, and exchange-correlation (XC) energies:

$$\mu = V(\mathbf{r}) + \frac{\delta F_{\text{HK}}[\rho]}{\delta\rho(\mathbf{r})}. \tag{72}$$

Thus, the Thomas–Fermi BEC equation takes the new form:

$$|\psi|^2 \cong \frac{1}{g}\left[\frac{\delta F_{\text{HK}}[\rho]}{\delta\rho(\mathbf{r})}\right]_{V(\mathbf{r})}, \tag{73}$$

which represents *the second* BEC–DFT *connection*, so to speak, because it represents another level of bosonic–fermionic coupling, as long as the Hohenberg–Kohn functional is associated with electrons and the rest of the quantities represent the bosonic properties. Of course, one can also formulate the bosonic version of the Hohenberg–Kohn functional [41–44].

However, the third working BEC–DFT connection can be established by employing the inter-bosonic average that successively gives [42]:

$$\begin{aligned}
\int d\mathbf{r}|\psi(\mathbf{r})|^4 &\cong \frac{1}{g^2}\int d\mathbf{r}[\mu - V(\mathbf{r})]^2\\
&= \frac{1}{g^2}\int d\mathbf{r}\left[\nabla_\rho F_{\mathrm{HK}}\right]^2\\
&= \frac{1}{g^2}\left\{\int d\mathbf{r}\nabla_\rho\left[(\nabla_\rho F_{\mathrm{HK}})F_{\mathrm{HK}}\right] - \int d\mathbf{r}F_{\mathrm{HK}}\left[\nabla_\rho^2 F_{\mathrm{HK}}\right]\right\}\\
&= \frac{1}{g^2}\int d\mathbf{r}\frac{\mathrm{d}}{d\mathbf{r}}\underbrace{\left(\frac{d\mathbf{r}}{\mathrm{d}\rho}\right)}_{\sim\frac{V^2}{N}}\left[(\nabla_\rho F_{\mathrm{HK}})F_{\mathrm{HK}}\right] - \frac{1}{g^2}\int d\mathbf{r}F_{\mathrm{HK}}\left[\nabla_\rho^2 F_{\mathrm{HK}}\right]\\
&= \underbrace{\frac{V^2}{Ng^2}\int_\infty \mathrm{d}\left[(\nabla_\rho F_{\mathrm{HK}})F_{\mathrm{HK}}\right]}_{\substack{\downarrow\\ 0}} - \frac{F_{\mathrm{HK}}}{g^2}\underbrace{\int d\mathbf{r}\frac{\delta^2 F_{\mathrm{HK}}}{\delta\rho^2(\mathbf{r})}}_{\sim\eta},
\end{aligned}$$

$$\int d\mathbf{r}|\psi(\mathbf{r})|^4 \cong -\frac{F_{\mathrm{HK}}\eta}{g^2}. \tag{74}$$

Whereas in the above-mentioned relationships the chemical hardness [54] was recognized through its local integrated version [55]:

$$\eta = \int d\mathbf{r}\eta(\mathbf{r})f(\mathbf{r}), \tag{75}$$

$$\eta(\mathbf{r}) = \frac{\delta\mu}{\delta\rho(\mathbf{r})} = \frac{\delta^2 F_{\mathrm{HK}}[\rho]}{\delta\rho^2(\mathbf{r})}, \tag{76}$$

here it adapted from the more general kernel version [56],

$$\eta(\mathbf{r},\mathbf{r}') = \frac{\delta^2 F_{\mathrm{HK}}[\rho]}{\delta\rho(\mathbf{r})\delta\rho(\mathbf{r}')}, \tag{77}$$

while using (ρ, V) as an independent variable alongside the celebrated (frontier) Fukui function $f(\mathbf{r})$, defined as

$$f(\mathbf{r}) = \left(\frac{\delta\mu}{\delta V(\mathbf{r})}\right)_N = \left(\frac{\delta\rho(\mathbf{r})}{\delta N}\right)_{V(\mathbf{r})}. \tag{78}$$

These BEC–DFT connections, which again represent a sort of fermionic–bosonic mixture, will be of primary use in emphasizing the difference BEC phenomena make in chemical bonding and to what degree they are relevant, as will be revealed in the next section.

4 BEC Levels of Chemical Bonding

4.1 Bondons: Particle Quantification of the Chemical Bonding Field

The chemical bond, perhaps the greatest challenge in theoretical chemistry, has generated many inspiring theses over the years, although none are definitive.

Few of the most preeminent scientists accept the orbital-based explanation of electronic pairing, in valence shells of atoms and molecules, rooted in the hybridization concept [57] that was then extended to valence-shell electron-pair repulsion (VSEPR) [58]. Alternatively, when electronic density was considered, the atoms-in-molecule paradigms were formulated through the geometrical partition of forces by Berlin [59], in terms of core, bonding, and lone-pair lodges by Daudel [60], and by the zero local flux in the gradient field of the density $\nabla\rho$ by Bader [61], until the most recent employment of the chemical action functional in bonding [44].

However, all of these approaches do not depart significantly from the undulatory nature of electronic motion in bonding, either by direct wave-function consideration or through its probabilistic nature manifested by electronic density (for that is still considered a condensed—observable version—of the undulatory manifestation of electrons). In other words, while passing from the Lewis point-like ansatz to the undulatory modeling of electrons in bonding, the reverse route is still missing, in terms of an analytical formulation. Only recently was the first attempt formulated, based on the broken-symmetry approach of the Schrödinger's Lagrangean with the electronegativity-chemical hardness parabolic energy dependency, showing that a systematical quest for the creation of particles from chemical bonding fields is possible [62].

Following this line, a step forward considers the gauge transformation of the electronic wave function and spinor over the de Broglie–Bohm augmented nonrelativistic and relativistic quantum pictures of the Schrödinger and Dirac electronic (chemical) fields, respectively. Consequently, the reality of the chemical field in bonding within either framework is expected in providing the corresponding bondonic particle with its respective mass and velocity in a full quantized form [63, 64].

The bondon is the quantum particle corresponding to superimposed electronic pairing effects or distribution in chemical bonds; accordingly, through the values of its mass and velocity, it may be possible to identify the type of bonding (in particular) the particle exhibits and to characterize its electronic behavior in bonding (in general).

The search for bondons proceeds according to the following algorithm [44, 63, 64]:

1. Consider the de Broglie–Bohm electronic wave-function/spinor Ψ_0 formulation of the associated quantum Schrödinger/Dirac equation of motion [65–68]:

$$\Psi_0(t,x) = R(t,x)\exp\left(i\frac{S(t,x)}{\hbar}\right), \tag{79}$$

with the R-amplitude and S-phase action factors given, respectively, as

$$R(t,x) = \sqrt{\Psi_0(t,x)^2} = \rho^{1/2}(x), \tag{80}$$

$$S(t,x) = px - Et, \tag{81}$$

in terms of electronic density ρ, momentum p, total energy E, and time-space (t, x) coordinates, without spin.

On the other side, written within the de Broglie–Bohm framework, the spinor solution looks like [69]:

$$\overleftrightarrow{\Psi}_0 = \frac{1}{\sqrt{2}}R(t,x)\begin{bmatrix}\varphi\\ \phi\end{bmatrix} = \frac{1}{\sqrt{2}}R(t,x)\begin{bmatrix}\exp\left\{\frac{i}{\hbar}[S(t,x)+s]\right\}\\ \exp\left\{-\frac{i}{\hbar}[S(t,x)+s]\right\}\end{bmatrix}, \quad s = \pm\frac{1}{2}, \tag{82}$$

which from consistency satisfies the necessary electronic density condition:

$$\overleftrightarrow{\Psi}_0^*\overleftrightarrow{\Psi}_0 = R^*R = \rho \tag{83}$$

2. Check for the recovery of the charge current conservation law:

$$\frac{\partial\rho}{\partial t} + \nabla\vec{j} = 0, \tag{84}$$

which assures the circulatory nature of the electronic fields under study.

3. Recognize the quantum potential V_{qua} and its equation,

$$V_{\text{qua}} = -\frac{\hbar^2}{2m}\frac{\nabla^2 R}{R}, \tag{85}$$

if it eventually appears.

4. Reload the electronic wave function/spinor under the augmented U(1) or SU(2) group form:

$$\Psi_G(t,x) = \Psi_0(t,x)\exp\left(\frac{i}{\hbar}\frac{e}{c}\aleph(t,x)\right), \tag{86}$$

with the standard abbreviation $e = e_0^2/4\pi\varepsilon_0$ in terms of the chemical field $\aleph$ considered as inverse of the fine-structure order:

$$\aleph_0 = \frac{\hbar c}{e} \sim 137.03599976\left[\frac{\text{Joule}\times\text{meter}}{\text{Coulomb}}\right], \tag{87}$$

which is upper bounded, in principle, by the atomic number of the ultimate chemical stable element ($Z = 137$). The quantity $\aleph_0$ plays a crucial role in chemical bonding, where the energies involved are on the order of 10^{-19} J (electron-volts)!

Nevertheless, to establish the physical significance of such chemical bonding quanta, one can proceed with the chain equivalences:

$$\aleph_B \sim \frac{\text{energy} \times \text{distance}}{\text{charge}} \sim \frac{\left(\text{charge} \times \begin{array}{c}\text{potential}\\ \text{difference}\end{array}\right) \times \text{distance}}{\text{charge}} \sim \left(\begin{array}{c}\text{potential}\\ \text{difference}\end{array}\right) \times \text{distance} \tag{88}$$

revealing that the chemical bonding field carries *bondons* with unit quanta $\hbar c/e$ along the distance of bonding within the potential gap of stability or by tunneling the potential barrier of encountered bonding attractors.

5. Rewrite the quantum wave-function/spinor equation with the group object Ψ_G while separating the terms containing the real and imaginary $\aleph$ chemical field contributions.
6. Identify the chemical field charge current and term within the actual group transformation context.
7. Establish the global/local gauge transformations that resemble the de Broglie–Bohm wave-function/spinor ansatz Ψ_0 of steps (1)–(3).
8. Impose invariant conditions for the Ψ_G wave function on the pattern quantum equation, according to the Ψ_0 wave-function/spinor action of steps (1)–(3).
9. Establish the $\aleph$ specific chemical field equations.
10. Solve the system of $\aleph$ chemical field equations.
11. Assess the stationary chemical field:

$$\frac{\partial \aleph}{\partial t} \equiv \partial_t \aleph = 0, \tag{89}$$

which is the case in chemical bonds at equilibrium (ground-state condition); this simplifies the quest for the solution to chemical field $\aleph$.

12. The manifested bondonic chemical field $\aleph_{\text{bondon}}$ is eventually identified along the bonding distance (or space).
13. Check the eventual charge flux condition of Bader within the vanishing chemical bonding field [61]:

$$\aleph_B = 0 \Leftrightarrow \nabla \rho = 0 \tag{90}$$

14. Employ the Heisenberg time–energy relaxation–saturation relationship through the kinetic energy of electrons in bonding to yield their velocity:

$$v = \sqrt{\frac{2T}{m}} \sim \sqrt{\frac{2}{m}\frac{\hbar}{t}} \tag{91}$$

15. Equate the bondonic chemical bond field with the chemical field quanta $\aleph_0$ to obtain the bondons' mass:

$$\aleph_B(m_B) = \aleph_0. \tag{92}$$

Once this algorithm is unfolded, either within for nonrelativistic or relativistic electronic motion, the need for bondon manifestation remains, while the allied proprieties are summarized next.

By performing the above algorithm, either by Schrodinger and Dirac quantum analysis, one obtains the working expression for the bondonic mass [63]:

$$m_B = \frac{\hbar^2}{2}\frac{(2\pi n+1)^2}{E_{\text{bond}}X_{\text{bond}}^2} \quad n = 0, 1, 2, \ldots \tag{93}$$

which is more practical than the traditional characterization of bonding types in terms of length and energy of bonding; it may further assume the numerical ground-state ratio form:

$$\varsigma_m = \frac{m_B}{m_0} = \frac{87.8603}{(E_{\text{bond}}[\text{kcal/mol}])\left(X_{\text{bond}}[\overset{0}{A}]\right)^2}, \tag{94}$$

when the available bonding energy and length are considered (as is the custom for chemical information) in kcal/mol and Angstroms, respectively. Note that expressing the bondon's mass in terms of bond energy implies the inclusion of the electronic pairing effect existence, without the constraint that the bonding pair may accumulate in the internuclear region [59].

Equally, the quantified bondon-to-light velocity ratio is determined to be [63]:

$$\frac{v_B}{c} = \frac{1}{\sqrt{1+\frac{1}{64\pi^2}\frac{\hbar^2c^2(2\pi n+1)^4}{E_{\text{bond}}^2X_{\text{bond}}^2}}} \quad n = 0, 1, 2\ldots \tag{95}$$

or numerically in the bonding ground state:

$$\varsigma_v = \frac{v_B}{c} = \frac{100}{\sqrt{1+\frac{3.27817\times10^6}{(E_{\text{bond}}[\text{kcal/mol}])^2\left(X_{\text{bond}}[\overset{0}{A}]\right)^2}}}[\%], \tag{96}$$

while it carries the electrical charge under the quantified form [63]:

$$e_B = \frac{4\pi\hbar c}{137.036}\frac{1}{\sqrt{1+\frac{64\pi^2E_{\text{bond}}^2X_{\text{bond}}^2}{\hbar^2c^2(2\pi n+1)^4}}}, \quad n = 0, 1, 2\ldots \tag{97}$$

employed towards the working ratio between the bondonic and electronic charges in the ground state of bonding:

$$\varsigma_e = \frac{e_B}{e} \sim \frac{1}{32\pi} \frac{\left(E_{\text{bond}}[\text{kcal/mol}]\right)\left(X_{\text{bond}}\left[\overset{0}{A}\right]\right)}{\sqrt{3.27817 \times 10^3}}. \tag{98}$$

With these quantities, the predicted lifetime of corresponding bondons, obtained from the working expressions for bondonic mass and velocity working through the basic time–energy Heisenberg relationship, is here restrained at the level of kinetic energy only for the bondonic particle; thus, one yields the subsequent analytical forms:

$$\begin{aligned} t_B &= \frac{\hbar}{T_B} = \frac{2\hbar}{m_B v_B^2} = \frac{2\hbar}{(m_0\varsigma_m)\left(c\varsigma_v \times 10^{-2}\right)^2} = \frac{\hbar}{m_0 c^2}\frac{2\times 10^4}{\varsigma_m \varsigma_v^2} \\ &= \frac{0.0257618}{\varsigma_m \varsigma_v^2} \times 10^{-15}[s]_{\text{SI}} \end{aligned} \tag{99}$$

Note that defining the bondonic lifetime using the last equation is the most adequate approach because it involves the basic bondonic (particle!) information, mass and velocity; instead, when directly evaluating the bondonic lifetime using only the bonding energy, one addresses the working formula:

$$t_{\text{bond}} = \frac{\hbar}{E_{\text{bond}}} = \frac{1.51787}{E_{\text{bond}}[\text{kcal/mol}]} \times 10^{-14}[s]_{\text{SI}}, \tag{100}$$

which usually produces values that are at least one order of magnitude lower than those reported in Table 1.

However, by calculating the special bonding length-energy properties, as reported in Table 2, one may note that the situation of the bondon having the same charge as the electron is quite improbable, at least for common chemical bonds, because in such a case the bondon velocity will approach the velocity of light (and almost no mass, i.e., a mass that continuously decreases as the bonding energy decreases and the bonding length increases). This is natural because a longer distance must be spanned by lowering the binding energy though carrying the same unit charge as an electron while it is transmitted at the same relativistic velocity! Such behavior may be regarded as what is currently referred to as *zitterbewegung* (trembling in motion) phenomena, here at the bondonic level. However, one records the systematic increase in the bondonic lifetime as being observable in the femtosecond regime for increasing bond length and decreasing bonding energy—under the condition that the chemical bonding itself still exists for certain $\{X_{\text{bond}}, E_{\text{bond}}\}$ combinations. On the other hand, the current situation is that in which the bondon will weigh as much as one electron (see the Table 1); nevertheless, it is accompanied by a quite reasonable chemical bonding length and energy information that can be carried at a small fraction of the

Table 1 Ratios for the bondon-to-electron mass and charge and for the bondon-to-light velocity, along the associated bondonic lifetime for typical chemical bonds in terms of their basic characteristics, such as the bond length and energy [70, 71], by employing the basic formulas, Eqs. (93)–(99), for the ground states, respectively [63]

Bond type	X_{bond} (Å)	E_{bond} (kcal/mol)	$\varsigma_m = \frac{m_B}{m_0}$	$\varsigma_v = \frac{v_B}{c}[\%]$	$\varsigma_e = \frac{e_B}{e}[\times 10^3]$	$t_B[\times 10^{15}]$ (s)
H–H	0.60	104.2	2.34219	3.451	0.3435	9.236
C–C	1.54	81.2	0.45624	6.890	0.687	11.894
C–C[a]	1.54	170.9	0.21678	14.385	1.446	5.743
C=C	1.34	147	0.33286	10.816	1.082	6.616
C≡C	1.20	194	0.31451	12.753	1.279	5.037
N≡N	1.10	225	0.32272	13.544	1.36	4.352
O=O	1.10	118.4	0.61327	7.175	0.716	8.160
F–F	1.28	37.6	1.42621	2.657	0.264	25.582
Cl–Cl	1.98	58	0.3864	6.330	0.631	16.639
I–I	2.66	36.1	0.3440	5.296	0.528	26.701
C–H	1.09	99.2	0.7455	5.961	0.594	9.724
N–H	1.02	93.4	0.9042	5.254	0.523	10.32
O–H	0.96	110.6	0.8620	5.854	0.583	8.721
C–O	1.42	82	0.5314	6.418	0.64	11.771
C=O[b]	1.21	166	0.3615	11.026	1.104	5.862
C=O[c]	1.15	191.6	0.3467	12.081	1.211	5.091
C–Cl	1.76	78	0.3636	7.560	0.754	12.394
C–Br	1.91	68	0.3542	7.155	0.714	14.208
C–I	2.10	51	0.3906	5.905	0.588	18.9131

[a]In diamond
[b]In CH_2O
[c]In O=C=O

velocity of light, however, with very small charge as well. Nevertheless, the discovered bonding energy–length relationship presented in Table 2, based on the above bondonic charge equation, namely [63],

$$E_{\text{bond}}[\text{kcal/mol}] \times X_{\text{bond}}[\overset{0}{A}] = 182{,}019, \tag{101}$$

should be used in setting appropriate experimental conditions in which the bondon particle B may be observed as carrying unit electronic charge though with almost zero mass.

In this way, *the bondon is affirmed to be a special particle in nature, which behaves like an electron with respect to charge, like a photon with respect to velocity and like a neutrino with respect to mass, while having an observable (at least as femtosecond) lifetime for nanosystems having chemical bond lengths and energies in the range of hundreds of Angstroms and thousands of kcal/mol, respectively* (see Fig. 1)!

Such peculiar nature of the bondon as the quantum particle of chemical bonding, the central theme of chemistry, is not as surprising when noting that chemistry

Table 2 Predicted basic values for bonding energy and length, along the associated bondonic lifetime and a fraction of the velocity of light for a system featuring unity ratios of bondonic mass and charge, respecting electronic values, by employing the basic formulas Eqs. (93)–(101) [63]

X_{bond} (Å)	E_{bond} (kcal/mol)	$t_{\mathcal{B}}[\times 10^{15}]$ (s)	$\varsigma_v = \frac{v_{\mathcal{B}}}{c}[\%]$	$\varsigma_m = \frac{m_{\mathcal{B}}}{m_0}$	$\varsigma_e = \frac{e_{\mathcal{B}}}{e}$
1	**87.86**	10.966	4.84691	1	0.4827×10^{-3}
1	**182019**	53.376	99.9951	4.82699×10^{-4}	1
10	**18201.9**	533.76	99.9951	4.82699×10^{-5}	1
100	**1820.19**	5337.56	99.9951	4.82699×10^{-6}	1

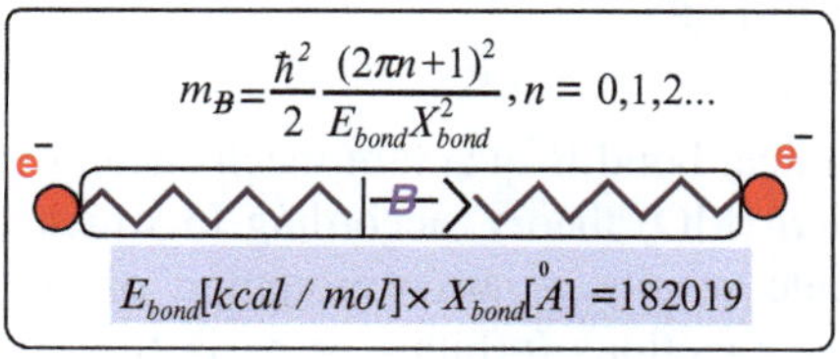

Fig. 1 The bondon as the quantum particle of chemical bonding interactions alongside its quantified mass and the Heisenberg-like relationship [72] given by Eqs. (93) and (101), respectively [63]

Table 3 Phenomenological classification of chemical bonding types by bondonic (mass, velocity, charge, and lifetime) properties borrowed from Table 1; the symbols used are > and > > for 'high' and 'very high' values, respectively; < and < < for 'low' and 'very low' values, respectively; ~ and ~ > for 'moderate' and 'moderate high and almost equal' values in their class of bonding, respectively [63]

Chemical bond	Property			
	ς_m	ς_v	ς_e	$t_{\mathcal{B}}$
Covalence	>>	<<	<<	>>
Multiple bonds	<	>	>	<
Metallic	<<	>	>	<
Ionic	~>	~	~	~>

seems to need both a particle view (such as that offered by relativity) and a wave view (such as that offered by quantum mechanics), although currently these two physics theories are not yet fully compatible with one another or even fully coherent internally. Perhaps the concept of "bondons" will help to improve the situation for all those concerned by its further conceptual applications.

The mass, velocity, charge, and lifetime properties of the bondons were employed to analyze some typical chemical bonds (see Table 1), thus revealing a sort of fuzzy classification of chemical bonding types in terms of the bondon-to-electron mass and charge ratios ς_m and ς_e and of the bondon-to-light velocity percent ratio ς_v, along with the bondonic observable life time, $t_{\mathcal{B}}$, as summarized in Table 3.

These rules are expected to be further refined by considering the new paradigms of special relativity in computing bondon velocities, especially within modern algebraic chemistry [73]. However, because the bondonic masses of chemical bonding ground states seem untouched by the Dirac relativistic considerations over the Schrödinger picture, it is expected that their analytical values may make a difference among the various types of compounds, while their experimental detection is hoped to be some day achieved.

4.2 General DFT–BEC Equations of Homopolar Chemical Bonding

Traditionally, the chemical bond is quantum mechanically approached by the so-called *molecular orbital* (MO) theory, according to which the pairing of atomic electrons in a molecule results as an overlapping or superposition of mono-electronic atomic wave functions (orbitals or basis functions, usually of *hydrogen-like* atoms because these are known analytically, i.e., *Slater-type orbitals*, though other choices are possible, such as *Gaussian functions* from standard *basis sets*) through the linear molecular combinations (orbitals):

$$\psi_i = c_{1i}\phi_1 + c_{2i}\phi_2 + \ldots + c_{Ni}\phi_N = \sum_{j=1}^{N} c_{ji}\phi_j, \tag{102}$$

as introduced in 1929 by Sir John Lennard–Jones [74]. The variational (or Hartree–Fock) procedure is used to determine the coefficients of the expansion and thus the molecular orbitals themselves [75–80].

The He–He interaction, which is too weak to produce chemical bonding and even resists solidification under a normal pressure below 10 K, has offered fruitful speculation regarding the existence of superfluid He, the prototype for the forthcoming Bose–Einstein condensation phenomena at the potential level (Fig. 2).

The special behavior of He_2 is motivated by the interparticle potential, see Fig. 2, which has the analytical expression [42, 44, 81]:

$$V(r) = \left(5.67 \times 10^6\right) \exp\left[-21.5\frac{r}{\sigma}\right] - 1.08\left(\frac{\sigma}{r}\right)^6 \ldots [^{\mathrm{o}}\mathbf{K}], \tag{103}$$

with the calculated average interparticle distance $\sigma = 4.64\,\overset{\mathrm{o}}{\mathrm{A}}$, thus displaying a special type of (Lennard–Jones) attractive potential. Nevertheless, while investigating the quantum localization of He atoms (or, more precisely, of their electrons) in bonding over, for example, a localization distance Δx that is relatively small compared with the potential range, one has the energy uncertainty (in Boltzmann's constant units of temperature) [45]:

$$\Delta E \cong \frac{1}{2m_{He}}\left(\frac{\hbar}{\Delta x}\right)^2 \xrightarrow{\Delta x \cong 0.5\,\overset{\mathrm{o}}{\mathrm{A}}} 10^{\mathrm{o}}\mathbf{K}, \tag{104}$$

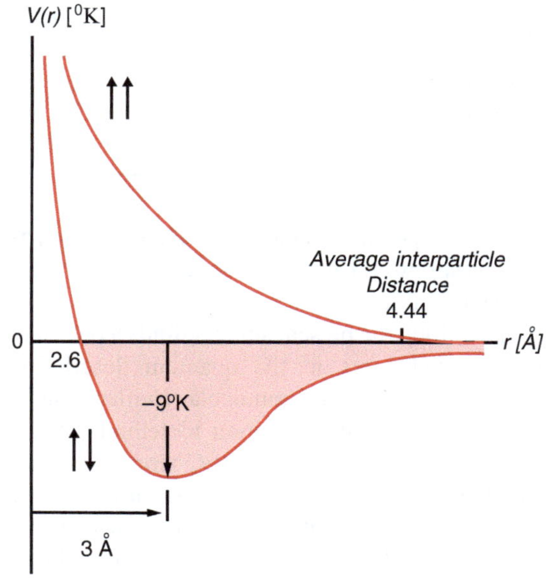

Fig. 2 Potential energy (in Kelvin) for homopolar He–He interaction as a function of interparticle separation distance [42, 44, 45]

which is comparable with the depths of the potential well, see Fig. 2; thus, electronic localization in He_2 is impossible, in close agreement with the molecular orbital prediction of zero bonding order for the chemical bond of He_2!

However, Bose–Einstein condensation (or BEC) has been primarily achieved in dilute gases of (weakly interacting) bosonic alkali atoms within external magnetic and/or optical traps (the external potential) with typical parameters such as 106 particles/condensate, an average density of approximately 1,012 cm^{-3} (i.e., the so-called thermodynamic limit with a practically infinite number of particles in an infinite space $N \to \infty, V \to \infty$), with a transition temperature on the order of 10^{-7} °K, though featuring a smaller density (by ten orders of magnitude) and temperature (by seven orders of magnitude) [16, 17, 82]. Such "extreme" behavior makes BEC systems suitable for being treated by the Thomas–Fermi approach, which is valid for systems with many particles, and by the N-many-body generalization as in the celebrated Density Functional Theory, especially for accurately treating atoms and molecules in general and chemical bonding in particular [42, 44, 69].

However, the resulting superfluid Landau theory of liquid helium, which has since been incorporated into the more general Bose–Einstein condensation phenomenology, should be reflected as a special contribution to the quantum description of chemical bonding has become apparent in some case and less so in others. This remark brings us to the forefront of the quest for being able to use bosonic condensation effects to describe chemical bonding, which will surely lead to a wide generalization for the current quantum approaches of bonding and reactivity due to the assumed fermionic–bosonic mixture of electronic basins in molecules and complex systems. Such an approach is encouraged by the observation that because chemical bonding is jointly represented by *bonding* and *antibonding* orbitals,

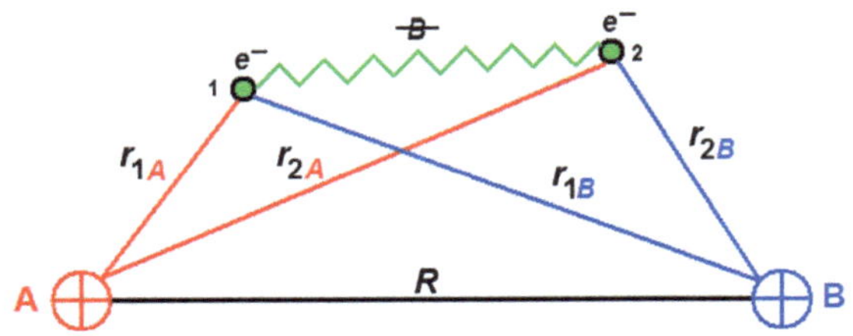

Fig. 3 Illustration of chemical bond formation by bondonic–bosonic contributions to interelectronic interactions

electronic spin arrangements can be interpreted as states in which electrons behave like *fermions* and *bosons*, respectively, coexisting as a mixture in the molecular space available.

Therefore, the present study unfolds a possible way of reconciling fermionic and bosonic behaviors in the quantum description of chemical bonding for the paradigmatic diatomic homopolar Heitler–London case [83].

Now, we are in a position to rethink chemical bonding, in its paradigmatic homopolar (A_2) version, in BEC terms, as was first accomplished by Heitler and London [83]; indeed, this view is currently becoming the pillar of the modern quantum (physical) approach to chemical bonding [84]. The starting point is to consider the Gross–Pitaevsky stationary Hamiltonian, which can be formally reconsidered for chemical bonding as being a mixed fermionic–bondonic (FB) system and cast as follows:

$$\hat{H}_{\mathrm{FB}} = \hat{H}_F + \hat{H}_B, \tag{105}$$

$$\hat{H}_F = -\frac{\hbar^2}{2m_0}\nabla^2 + V(\mathbf{r}). \tag{106}$$

Figure 3 illustrates the newly developed quantum condensation paradigm of chemical bonding: the interelectronic interaction is replaced by the actual bondonic–bosonic (Hamiltonian) contribution, while the rest of the electron-nuclei and internuclear interactions are included in the conventional fermionic Hamiltonian. It is worth noting that a system considered to be chemically bonded requires the presence of two electrons, i.e., of at least one bondon, otherwise the bond is unstable (e.g., the hydrogenic molecular ion, H_2^+); in other words, the chemical bond is due not to the two-nucleus approach with a common electron but due to the two electrons in the two-nucleus framework that manifests quantum condensation rather than electrostatic repulsion; in this way, the bondonic paradigm replaces the exchange-correlation phenomenology with the bosonic condensation phenomenology.

Within this framework, for bonding electrons (with their mass m_0), one has the generic bosonic condensation Hamiltonian, rooted in the nonlinear Schrodinger or Gross–Pitaevsky equation (67),

$$\hat{H}_B = g|\psi|^2, \tag{107}$$

which accounts for bosonic formation (and eventually condensation) by the self-interaction coupling, which is here considered to characterize the bondons with the ground-state mass abstracted from Eq. (93) as follows [63]:

$$m_B = \frac{\hbar^2}{2} \frac{1}{E_{\text{bond}} R_{\text{bond}}^2}, \tag{108}$$

where $(E_{\text{bond}}, R_{\text{bond}})$ is the bonding equilibrium energy and distance. In other terms, the bondonic bosonic strength can be expressed as

$$g_B = \left.\frac{4\pi a_B \hbar^2}{m_B}\right|_{a_B \to R_{\text{bond}}} = 8\pi E_{\text{bond}} R_{\text{bond}}^3, \tag{109}$$

where the bonding volume and energy are the main factors affecting bonding strength. More importantly, due to the bondonic volume dependence, the bosonic self-interaction coupling can be further expressed by regaining the order parameter $\langle\psi(\mathbf{r})\rangle$ dependency by considering the cubic volume as the inscribing locus of the bonding so that we first have

$$R_{\text{bond}}^3 = V_{\text{bond}} = \frac{N_{\text{bondons}}}{\underset{\text{(bosons)}}{\rho_{\text{bodnons}}}(\mathbf{r})} \overset{\substack{\text{(simple}\\ \text{bonding:}\\ \text{one bondon)}}}{=} \frac{1}{|\langle\psi(\mathbf{r})\rangle|^2}, \tag{110}$$

and subsequently for the bosonic coupling

$$g_B = \frac{8\pi E_{\text{bond}}}{|\langle\psi(\mathbf{r})\rangle|^2}. \tag{111}$$

This formula represents the bonding-level reformulation for the inverse relationship between $|\langle\psi(\mathbf{r})\rangle|^2$ and g_B, as already noted from the above Thomas–Fermi approximation. However, for consistency, it is worth noting that the order parameter refers to the atomic gas bosonic property in the isolated condition, i.e., when it is treated only as a bosonic condensate alone; this acquires chemical bonding relevance when combined with the bonding energy in g_B.

Next, the expectation value of this bondonic–fermionic Hamiltonian is formed over the bondonic (bosonic) molecular wave function that arises from the superposition of the fermionic atomic wave functions (the so-called *build-in-bondonic* BB superposition of fermionic wave functions coming from the two atoms labeled *A* and *B* to better illustrate the combined electronic effects among them) [42, 44]

$$\underbrace{\psi_{\mathrm{BB}}(\mathbf{r})}_{\text{BONDONIC}} = \underbrace{c_A\psi_A(\mathbf{r}) + c_B\psi_B(\mathbf{r})}_{\text{FERMIONIC SUPERPOSITION}} \quad c_A, c_B \in \Re, \tag{112}$$

thus giving rise to the fermionic and bosonic contributions:

$$\begin{aligned}\langle \hat{H}_{\mathrm{FB}} \rangle &= \langle \hat{H}_F \rangle_{\substack{\text{FERMIONIC}\\ \text{SIDE}}} + \langle \hat{H}_B \rangle_{\substack{\text{BOSONIC}\\ \text{SIDE}}} \\ &= \frac{\int d\mathbf{r}(c_A\psi_A(\mathbf{r}) + c_B\psi_B(\mathbf{r}))^* \hat{H}_F (c_A\psi_A(\mathbf{r}) + c_B\psi_B(\mathbf{r})) + g\int d\mathbf{r}|\psi_{\mathrm{BB}}(\mathbf{r})|^4}{\int d\mathbf{r}(c_A\psi_A(\mathbf{r}) + c_B\psi_B(\mathbf{r}))^*(c_A\psi_A(\mathbf{r}) + c_B\psi_B(\mathbf{r}))}.\end{aligned} \tag{113}$$

The homo-polar framework of binding is nevertheless reflected in the following shortcuts for

- *intra-atomic*

$$H_{11} = \int d\mathbf{r}\psi_A^*(\mathbf{r})\hat{H}_F\psi_A(\mathbf{r}) = \int d\mathbf{r}\psi_B^*(\mathbf{r})\hat{H}_F\psi_B(\mathbf{r}), \tag{114}$$

- *interatomic*

$$H_{12} = \int d\mathbf{r}\psi_A^*(\mathbf{r})\hat{H}_F\psi_B(\mathbf{r}) = \int d\mathbf{r}\psi_B^*(\mathbf{r})\hat{H}_F\psi_A(\mathbf{r}), \tag{115}$$

- and *overlapping*

$$S = \int d\mathbf{r}\psi_A^*(\mathbf{r})\psi_B(\mathbf{r}) = \int d\mathbf{r}\psi_B^*(\mathbf{r})\psi_A(\mathbf{r}), \tag{116}$$

terms. However, by considering the above BEC–DFT expression for the ψ^4 integral, eq. (74), we further obtain the expression:

$$\langle \hat{H}_{\mathrm{FB}} \rangle \cong \frac{1}{c_A^2 + c_B^2 + 2c_Ac_BS}\left\{(c_A^2 + c_B^2)H_{11} + 2c_Ac_BH_{12} - \frac{F_{\mathrm{HK}}\eta_{\mathrm{Molec}}}{g_B}\right\}, \tag{117}$$

where bonding (bondonic) chemical hardness is now assimilated into the chemical hardness of the molecule itself, while for the Hohenberg–Kohn functional we can still employ the previous basic DFT Eq. (71), which is adapted as follows:

$$\begin{aligned} F_{\mathrm{HK}} &= \langle \hat{H}_{\mathrm{FB}} \rangle - C_A \\ &= \langle \hat{H}_{\mathrm{FB}} \rangle - \int d\mathbf{r} V(\mathbf{r}) \left[(c_A \psi_A(\mathbf{r}) + c_B \psi_A(\mathbf{r}))^* (c_A \psi_A(\mathbf{r}) + c_B \psi_A(\mathbf{r})) \right] \\ &= \langle \hat{H}_{\mathrm{FB}} \rangle - (c_A^2 + c_B^2) V_{11} - 2 c_A c_B V_{12}. \end{aligned} \tag{118}$$

In this case, we have introduced the natural potential integral notations paralleling the Coulombic and exchange contributions for electrons in bonding, namely,

$$V_{11} = \int d\mathbf{r} \psi_A^*(\mathbf{r}) V(\mathbf{r}) \psi_A(\mathbf{r}) = \int d\mathbf{r} \psi_B^*(\mathbf{r}) V(\mathbf{r}) \psi_B(\mathbf{r}), \tag{119}$$

$$V_{12} = \int d\mathbf{r} \psi_A^*(\mathbf{r}) V(\mathbf{r}) \psi_B(\mathbf{r}) = \int d\mathbf{r} \psi_B^*(\mathbf{r}) V(\mathbf{r}) \psi_A(\mathbf{r}). \tag{120}$$

With these, we obtain an algebraic equation for the fermionic–bondonic Hamiltonian expectation value:

$$\langle \hat{H}_{\mathrm{FB}} \rangle \cong \frac{(c_A^2 + c_B^2)\left(H_{11} + \frac{\eta_{\mathrm{Molec}}}{g_B} V_{11}\right) + 2 c_A c_B \left(H_{12} + \frac{\eta_{\mathrm{Molec}}}{g_B} V_{12}\right) - \langle \hat{H}_{\mathrm{FB}} \rangle \frac{\eta_{\mathrm{Molec}}}{g_B}}{c_A^2 + c_B^2 + 2 c_A c_B S}, \tag{121}$$

which can be conveniently rearranged as follows:

$$\langle \hat{H}_{\mathrm{FB}} \rangle \cong \frac{(c_A^2 + c_B^2)\left(H_{11} + \frac{\eta_{\mathrm{Molec}}}{g_B} V_{11}\right) + 2 c_A c_B \left(H_{12} + \frac{\eta_{\mathrm{Molec}}}{g_B} V_{12}\right)}{c_A^2 + c_B^2 + 2 c_A c_B S + \frac{\eta_{\mathrm{Molec}}}{g_B}}, \tag{122}$$

so that by comparing with the standard molecular orbital (MO) theory one can say that the BEC influence is "manifested" over MO binding on both the energetic (numerator) and normalization (denominator) sides. However, through the actual BEC–DFT connections, we succeed in distributing the BEC effects such that the fundamental nonlinear Schrödinger effects on the Gross–Pitaevsky equation are absorbed by the chemical hardness (as fermionic-electronic quantity) and ratio with respect to the self-bosonic (bondonic) coupling. Thus, the original nonlinear problem, which is otherwise unlikely solvable for nonzero potentials, is reformulated in an elegant manner within the traditional MO-binding problem, though corrected with additional terms dispersed throughout the formulation. Next, by unfolding the variational principle with respect to the MO coefficients [42, 44]:

$$\begin{cases} \frac{\partial}{\partial c_A} \langle \hat{H}_{\mathrm{FB}} \rangle = 0 \\ \frac{\partial}{\partial c_B} \langle \hat{H}_{\mathrm{FB}} \rangle = 0, \end{cases} \tag{123}$$

the symmetric system with respect to the (c_A, c_B) coefficients is established:

$$\begin{cases} c_A\left(H_{11}+\frac{\eta_{\text{Molec}}}{g_B}V_{11}-\langle\hat{H}_{\text{FB}}\rangle\right)+c_B\left(H_{12}+\frac{\eta_{\text{Molec}}}{g_B}V_{12}-\langle\hat{H}_{\text{FB}}\rangle S\right)=0 \\ c_A\left(H_{12}+\frac{\eta_{\text{Molec}}}{g_B}V_{12}-\langle\hat{H}_{\text{FB}}\rangle S\right)+c_B\left(H_{11}+\frac{\eta_{\text{Molec}}}{g_B}V_{11}-\langle\hat{H}_{\text{FB}}\rangle\right)=0 \end{cases} \tag{124}$$

whose secular determinant:

$$\begin{vmatrix} H_{11}+\frac{\eta_{\text{Molec}}}{g_B}V_{11}-\langle\hat{H}_{\text{FB}}\rangle & H_{12}+\frac{\eta_{\text{Molec}}}{g_B}V_{12}-\langle\hat{H}_{\text{FB}}\rangle S \\ H_{12}+\frac{\eta_{\text{Molec}}}{g_B}V_{12}-\langle\hat{H}_{\text{FB}}\rangle S & H_{11}+\frac{\eta_{\text{Molec}}}{g_B}V_{11}-\langle\hat{H}_{\text{FB}}\rangle \end{vmatrix} = 0, \tag{125}$$

leads to the double-solution equation [42, 44]:

$$H_{11}+\frac{\eta_{\text{Molec}}}{g_B}V_{11}-\langle\hat{H}_{\text{FB}}\rangle = \pm\left(H_{12}+\frac{\eta_{\text{Molec}}}{g_B}V_{12}-\langle\hat{H}_{\text{FB}}\rangle S\right). \tag{126}$$

The first choice corresponds to the $(+)$ sign in Eq. (126) and provides the *antibonding* energy and molecular wave function [42, 44]:

$$\begin{cases} \langle\hat{H}_{\text{FB}}\rangle^{\beta} \equiv E^{-} = \dfrac{H_{11}-H_{12}}{1-S}+\dfrac{\eta_{\text{Molec}}}{g_B}\dfrac{V_{11}-V_{12}}{1-S} \\ \psi_{\text{BB}}^{\beta}(\mathbf{r}) = \dfrac{1}{\sqrt{2-2S}}[\psi_A(\mathbf{r})-\psi_A(\mathbf{r})] \end{cases} \tag{127}$$

The second choice corresponds to the $(-)$ sign in Eq. (126) and produces the *bonding* energy and molecular wave function [42, 44]:

$$\begin{cases} \langle\hat{H}_{\text{FB}}\rangle^{\alpha} \equiv E^{+} = \dfrac{H_{11}+H_{12}}{1+S}+\dfrac{\eta_{\text{Molec}}}{g_B}\dfrac{V_{11}+V_{12}}{1+S} \\ \psi_{\text{BB}}^{\alpha}(\mathbf{r}) = \dfrac{1}{\sqrt{2+2S}}[\psi_A(\mathbf{r})+\psi_A(\mathbf{r})] \end{cases} \tag{128}$$

under the same condition of the conservation of the number of bondons in the system (the N-body bosonic condition), which here, however, is reduced at unity because a basic *single chemical bond* considered, i.e.,

$$1 = \int d\mathbf{r}[\psi_{\text{BB}}(\mathbf{r})]^2. \tag{129}$$

However, we would like to comment on and interpret the present BEC–DFT results on chemical bonding basic quantum-mechanically (orbital-molecular) theory. In this respect it is worth noting that [42, 44]

- The bosonic (bondonic)–fermionic mixture is assumed in chemical binding, i.e., the molecular orbital is viewed as representing the bondon formed by the fermionic electronic orbital contribution arising from the binding atoms;
- The observed bonding and antibonding energies from the bondonic–fermionic mixture of chemical bonding display the custom fermionic part plus an additional potential part modulated by the bondonic coupling and the fermionic chemical hardness of the bonding and molecule, respectively;
- The resulting bonding and antibonding molecular wave function shapes are essentially unmodified forms taken from classical MO theory, leading to the idea that for the same orbital there are fermionic and fermionic–bondonic forms that exist; however, they separated by the gaps $(\eta_{\mathrm{Molec}}/g_B)[(V_{11} \pm V_{12})/(1 \pm S)]$, respectively. Nevertheless, one may argue that according to the Hugenholtz–Pines theorem [85], an interacting Bose gas does not exhibit an energy gap in the case of repulsive interactions, with $g > 0$, though here we are addressing a Fermi–Bose mixture, so the gap under discussion will be between the pure fermionic and fermionic–bosonic systems; therefore, we will call it the *bosonic–fermionic degeneration* of chemical bonding;
- Regarding the position of the newly predicted energetic bosonic states, one can observe, following the same algebraic forms of the pure fermionic leading terms, that the bonding bosonic state will be shifted such that the total energy becomes more negative, while for the antibonding bosonic state the total energy becomes more positive—while being actually observable in the bosonic temperature range of a Bose–Einstein condensate, i.e., far below 10°[K], on the nano-Kelvin scale, in fact; however, conceptually, because the quantum mechanical calculations for chemical systems tacitly assume the 0°[K] conditions, both the lower fermionic–bosonic ground state as well as the higher valence (antibonding) state are relevant in rethinking the chemical reactivity with the aid of frontier-controlled orbitals.

Remarkably, it is worth pointing out that the bosonic nature of chemical bonding primarily depends on the degree to which the atoms involved in bonding display bosonic condensation when considered as a single gas having a nonzero order parameter $\langle\psi(\mathbf{r})\rangle \neq 0$ because the BEC side of the chemical bonding imply terms such as [42, 44]

$$\mathrm{BEC}_{\substack{\mathrm{CHEMICAL}\\ \mathrm{BONDING}}} = |\langle\psi(\mathbf{r})\rangle|^2 \frac{\eta_{\mathrm{Molec}}}{8\pi E_{\mathrm{bond}}} \frac{V_{11} \pm V_{12}}{1 \pm S}, \tag{130}$$

for the (+) bonding and (−) antibonding states of the chemical bonding phenomenon, respectively. However, it is also relevant that by replacing the BEC term in the bonding and antibonding energies of Eqs. (127) and (128) one obtain the following working equations:

$$E_{\text{bond}}^{\pm} = \frac{H_{11} \pm H_{12}}{1 \pm S} + |\langle\psi(\mathbf{r})\rangle|^2 \frac{\eta_{\text{Molec}}}{8\pi E_{\text{bond}}^{\pm}} \frac{V_{11} \pm V_{12}}{1 \pm S}, \tag{131}$$

which assume the quadratic form:

$$x = \Xi + \frac{\Theta}{x}, \tag{132}$$

with identifications:

$$\Xi \equiv \frac{H_{11} \pm H_{12}}{1 \pm S}, \tag{133}$$

$$\Theta \equiv |\langle\psi(\mathbf{r})\rangle|^2 \frac{\eta_{\text{Molec}}}{8\pi} \frac{V_{11} \pm V_{12}}{1 \pm S}. \tag{134}$$

The basic solutions to Eq. (132) assume the form:

$$x_{1,2} = \frac{1}{2}\left[\Xi \pm \left(\Xi^2 + 4\Theta\right)^{1/2}\right], \tag{135}$$

and provide, in the limit of small bosonic interaction, i.e., $\Theta \sim \langle\psi\rangle \to 0$, as is the case for bondons in chemical bonding, the following working expressions:

$$x_1 = -\frac{\Theta}{\Xi}, \tag{136}$$

$$x_2 = \Xi + \frac{\Theta}{\Xi}. \tag{137}$$

These account for the actual BEC–DFT twofold bonding and antibonding energies [42, 44]:

$$E_{\text{bond-BEC-I}}^{\pm} = -|\langle\psi(\mathbf{r})\rangle|^2 \frac{\eta_{\text{Molec}}}{8\pi} \frac{V_{11} \pm V_{12}}{H_{11} \pm H_{12}}, \tag{138}$$

$$E_{\text{bond-BEC-II}}^{\pm} = \frac{H_{11} \pm H_{12}}{1 \pm S} + |\langle\psi(\mathbf{r})\rangle|^2 \frac{\eta_{\text{Molec}}}{8\pi} \frac{V_{11} \pm V_{12}}{H_{11} \pm H_{12}}, \tag{139}$$

which can be appropriately called *physical- and chemical- bonding condensates*, respectively, due to the specific dependencies they incorporate.

The results of Eqs. (138) and (139) considerably enlarge (and enrich) the chemical bonding paradigm—easily recognized in the first term of Eq. (139), where the BEC phenomenology is completely neglected—to include the condensation and ordering phenomena, especially those associated with ^{4}He or alkali gases [86]. Equally relevant is the presence of the molecular chemical hardness in the numerator of the BEC chemical bonding term Eq. (130), which further supports the condensation phenomenology due to its maximum hardness principle for a stable molecule [87], i.e., the higher the hardness, the higher the stability of the chemical system.

4.3 Bonding DFT–BEC Energies for Homopolar Molecules

The atomic systems of units, which that correspond to atomic dimensions, are adopted and eliminate the physical constants from the Schrödinger equation; specifically, the length and energy units become the first Bohr radius $a_0 = 0.529\,\mathring{A}$ and the interaction energy between two units of charge separated by the Bohr radius (Hartree) $\varepsilon_0 = 27.246\,\text{eV}$.

For homopolar chemical binding, the fermionic diatomic potential, in atomic units, is adapted from Fig. 3 to the working form:

$$V(A,B) = -\frac{2}{r_A} - \frac{2}{r_B} + \frac{1}{R}. \tag{140}$$

Note that the actual quantum condensation paradigm of chemical bonding views the interelectronic interaction as a *condensed bondon* superimposed on the two independent/equivalent electron-nuclei fermionic interactions.

By introducing the specific *Coulombic* integral [88, 89]:

$$I = \int d\tau \psi_A^*(r_A) \frac{1}{r_B} \psi_A(r_A) = \left\langle \psi_A \left| \frac{1}{r_B} \right| \psi_A \right\rangle, \tag{141}$$

and the *exchange:*

$$J = \int d\tau \psi_A^*(r_A) \frac{1}{r_A} \psi_B(r_B) = \left\langle \psi_A \left| \frac{1}{r_A} \right| \psi_B \right\rangle, \tag{142}$$

one has for the associate fermionic energy integrals Eq. (114) the representative form:

$$\begin{aligned} H_{11} &= \int d\tau \psi_A^*(r_A) \left[-\frac{1}{2}\nabla^2 - \frac{2}{r_A} - \frac{2}{r_B} + \frac{1}{R} \right] \psi_A(r_A) \\ &= \left\langle \psi_A \left| -\frac{1}{2}\nabla^2 - \frac{1}{r_A} \right| \psi_A \right\rangle - \left\langle \psi_A \left| \frac{1}{r_A} \right| \psi_A \right\rangle + \left\langle \psi_A \left| \frac{1}{R} \right| \psi_A \right\rangle - \left\langle \psi_A \left| \frac{2}{r_B} \right| \psi_A \right\rangle \\ &= E_0 - U + \frac{N_0}{R} - 2I, \end{aligned} \tag{143}$$

by considering the atomic DFT-type normalization relationships for the valence electrons:

$$\begin{cases} \langle\psi_A|\psi_A\rangle = \langle\psi_B|\psi_B\rangle = N_0 \\ N_0 = 1, 2. \end{cases} \tag{144}$$

the *atomic* Eigen-energy equation in the ground state:

$$\left(-\frac{1}{2}\nabla^2 - \frac{1}{r_A}\right)|\psi_A\rangle = E_0|\psi_A\rangle, \tag{145}$$

and the self-interaction (a sort of internal) energy:

$$U = \left\langle \psi_A \left| \frac{1}{r_A} \right| \psi_A \right\rangle. \tag{146}$$

Analogously, for the integral Eq. (115), one successively yields

$$\begin{aligned} H_{12} &= \int d\tau \psi_A^*(r_A)\left[-\frac{1}{2}\nabla^2 - \frac{2}{r_A} - \frac{2}{r_B} + \frac{1}{R}\right]\psi_B(r_B) \\ &= \underbrace{\langle\psi_A|\left[-\frac{1}{2}\nabla^2 - \frac{1}{r_B}\right]|\psi_B\rangle}_{E_0|\psi_B\rangle} - \left\langle \psi_A \left| \frac{1}{r_B} \right| \psi_B \right\rangle + \frac{1}{R}\underbrace{\langle\psi_A|\psi_B\rangle}_{S} - \left\langle \psi_A \left| \frac{2}{r_A} \right| \psi_B \right\rangle \\ &= E_0 S + \frac{S}{R} - 3J, \end{aligned} \tag{147}$$

while recognizing the influence of the overlap integral Eq. (116)

$$\langle\psi_A|\psi_B\rangle = \langle\psi_B|\psi_A\rangle = S, \tag{148}$$

and the atomic energy Eq. (145), which is derived from the B-atomic contribution.

Additionally, the newly developed integrals Eqs. (119) and (120) unfold, following the same arguments, respectively:

$$\begin{aligned} V_{11} &= \int d\tau \psi_A^*(r_A)\left[-\frac{2}{r_A} - \frac{2}{r_B} + \frac{1}{R}\right]\psi_A(r_A) \\ &= -2\underbrace{\left\langle \psi_A \left| \frac{1}{r_A} \right| \psi_A \right\rangle}_{U} - 2\underbrace{\left\langle \psi_A \left| \frac{1}{r_B} \right| \psi_A \right\rangle}_{I} + \frac{\langle\psi_A|\psi_A\rangle}{R} \\ &= -2U - 2I + \frac{N_0}{R}, \end{aligned} \tag{149}$$

$$V_{12} = \int d\tau \psi_A^*(r_A)\left[-\frac{2}{r_A} - \frac{2}{r_B} + \frac{1}{R}\right]\psi_B(r_B)$$
$$= -2\underbrace{\left\langle \psi_A \left| \frac{1}{r_A} \right| \psi_B \right\rangle}_{J} - 2\underbrace{\left\langle \psi_A \left| \frac{1}{r_B} \right| \psi_B \right\rangle}_{J} + \frac{S}{R}$$
$$= -4J + \frac{S}{R}. \tag{150}$$

With these, one employs the results:

$$H_{11} + H_{12} = E_0(1+S) + \frac{N_0 + S}{R} - U - 2I - 3J, \tag{151}$$

$$H_{11} - H_{12} = E_0(1-S) + \frac{N_0 - S}{R} - U - 2I + 3J, \tag{152}$$

$$V_{11} + V_{12} = -2U - 2I - 4J + \frac{N_0 + S}{R}, \tag{153}$$

$$V_{11} - V_{12} = -2U - 2I + 4J + \frac{N_0 - S}{R} \tag{154}$$

to re-express the physical and chemical bonding condensates' equations (138) and (139), respectively, as

$$E^{\pm}_{\text{bond-BEC-I}} = -|\langle\psi(\mathbf{r})\rangle|^2 \frac{\eta_{\text{Molec}}}{8\pi} \frac{(-2U - 2I \mp 4J)R + N_0 \pm S}{[E_0(1 \pm S) - U - 2I \mp 3J]R + N_0 \pm S}, \tag{155}$$

$$E^{\pm}_{\text{bond-BEC-II}} = E_0 + \frac{N_0 \pm S}{(1 \pm S)R} - \frac{U + 2I \pm 3J}{1 \pm S} - E^{\pm}_{\text{bond-BEC-I}}. \tag{156}$$

Further evaluations of expressions (155) and (156) are possible by calculating the involved integrals using elliptic coordinates, see Fig. 4, i.e., by coordinate transformations:

$$\lambda = \frac{r_A + r_B}{R} \in [1, \infty), \tag{157}$$

$$\delta = \frac{r_A - r_B}{R} \in [-1, +1], \tag{158}$$

$$\varphi = \varphi_A = \varphi_B \in [0, 2\pi], \tag{159}$$

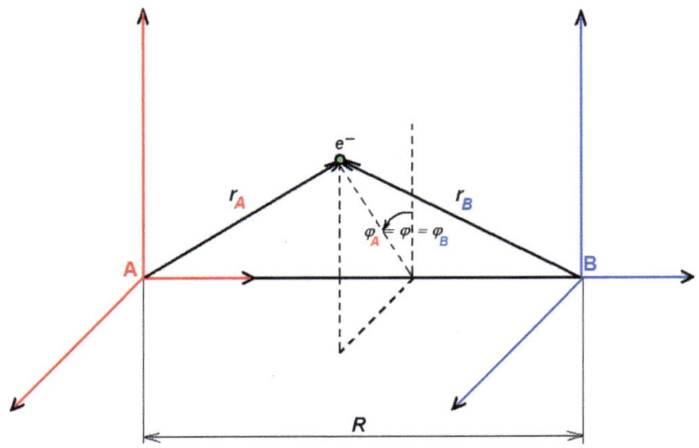

Fig. 4 Elliptic coordinate system for the bondonic equivalent electron "immersed" within a diatomic molecular system

producing the elementary volume

$$d\tau = \frac{R^3}{8}(\lambda^2 - \delta^2)d\lambda d\delta d\varphi, \tag{160}$$

and the inverse relationships

$$r_A = \frac{R(\lambda + \delta)}{2}, \tag{161}$$

$$r_B = \frac{R(\lambda - \delta)}{2}. \tag{162}$$

These geometric conditions are applied to the special 1s atomic orbitals, suitable for modeling the bosonic interactions as well, with semiclassical variational wave functions [90, 91] in atomic units:

$$\psi(r) = \sqrt{\frac{\gamma}{\pi}}\exp(-\alpha r), \tag{163}$$

with the variational parameters

$$\alpha = \xi,\ \gamma = N_0\xi^3,\ \xi = \frac{21 - 5N_0}{16}, \tag{164}$$

which features the following important properties:

- Integrates the atomic (or valence) number of electrons, according to a DFT-like normalization condition (144)

- Includes the orbital contribution Eq. (164), which recovers the 1s hydrogen wave function (when $N_0 = 1$)

Under these conditions, one may proceed with the analytical unfolding of the contributing integrals in Eqs. (155) and (156) as follows:

- The overlap integral Eq. (148) produces

$$
\begin{aligned}
S = \langle \psi_A | \psi_B \rangle &= \frac{\gamma}{\pi} \int e^{-\alpha(r_A + r_B)} \mathrm{d}\tau = \frac{\gamma}{\pi} \int e^{-\alpha R \lambda} \mathrm{d}\tau \\
&= \frac{\gamma R^3}{8\pi} \int_1^{\infty} \int_{-1}^{+1} \int_0^{2\pi} e^{-\alpha R \lambda} \left(\lambda^2 - \delta^2\right) \mathrm{d}\lambda \mathrm{d}\delta \mathrm{d}\varphi \\
&= \frac{\gamma R^3}{4} \left[\int_{-1}^{+1} \mathrm{d}\delta \int_1^{\infty} \lambda^2 e^{-\alpha R \lambda} \mathrm{d}\lambda - \int_{-1}^{+1} \delta^2 \mathrm{d}\delta \int_1^{\infty} e^{-\alpha R \lambda} \mathrm{d}\lambda \right] \\
&= \frac{\gamma R^3}{2} \left[\left(\frac{1}{\alpha R} + \frac{2}{\alpha^2 R^2} + \frac{2}{\alpha^3 R^3} \right) e^{-\alpha R} - \frac{1}{3\alpha R} e^{-\alpha R} \right] \\
&= \gamma \left(\frac{1}{\alpha^3} + \frac{R}{\alpha^2} + \frac{R^2}{3\alpha} \right) e^{-\alpha R} \\
&= N_0 \left(1 + \xi R + \frac{1}{3} \xi^2 R^2 \right) e^{-\xi R}. \qquad (165)
\end{aligned}
$$

- The exchange integral Eq. (142) analytically yields

$$
\begin{aligned}
J = \left\langle \psi_A \middle| \frac{1}{r_A} \middle| \psi_B \right\rangle &= \frac{\gamma}{\pi} \int \frac{e^{-\alpha(r_A + r_B)}}{r_A} \mathrm{d}\tau \\
&= \frac{\gamma R^3}{8\pi} \int_1^{\infty} \int_{-1}^{+1} \int_0^{2\pi} e^{-\alpha R \lambda} \left(\lambda^2 - \delta^2\right) \frac{2}{R(\lambda + \delta)} \mathrm{d}\lambda \mathrm{d}\delta \mathrm{d}\varphi \\
&= \frac{\gamma R^2}{2} \left[\int_{-1}^{+1} \mathrm{d}\delta \int_1^{\infty} \lambda e^{-\alpha R \lambda} \mathrm{d}\lambda - \int_{-1}^{+1} \delta \mathrm{d}\delta \int_1^{\infty} e^{-\alpha R \lambda} \mathrm{d}\lambda \right] \\
&= \gamma \left(R + \frac{1}{\alpha} \right) e^{-\alpha R} \\
&= N_0 \xi^2 (1 + \xi R) e^{-\xi R}. \qquad (166)
\end{aligned}
$$

- The Coulombic integral Eq. (141) equivalently becomes

$$
\begin{aligned}
I &= \left\langle \psi_A \middle| \frac{1}{r_B} \middle| \psi_A \right\rangle = \frac{\gamma}{\pi} \int \frac{e^{-2\alpha r_A}}{r_B} \mathrm{d}\tau \\
&= \frac{\gamma R^3}{8\pi} \int_1^{\infty} \int_{-1}^{+1} \int_0^{2\pi} e^{-\alpha R(\lambda+\delta)} \frac{2}{R(\lambda-\delta)} (\lambda^2 - \delta^2) \mathrm{d}\lambda \mathrm{d}\delta \mathrm{d}\varphi \\
&= \frac{\gamma R^2}{2} \int_1^{\infty} \int_{-1}^{+1} e^{-\alpha R(\lambda+\delta)} (\lambda + \delta) \mathrm{d}\lambda \mathrm{d}\delta \\
&= \frac{\gamma R^2}{2} \left[\left(\int_1^{\infty} \lambda e^{-\alpha R \lambda} \mathrm{d}\lambda \right) \left(\int_{-1}^{+1} e^{-\alpha R \delta} \mathrm{d}\delta \right) + \left(\int_1^{\infty} e^{-\alpha R \lambda} \mathrm{d}\lambda \right) \left(\int_{-1}^{+1} \delta e^{-\alpha R \delta} \mathrm{d}\delta \right) \right] \\
&= \frac{\gamma}{2} \left[-\frac{1}{\alpha} \left(1 + \frac{1}{\alpha} \right) \left(1 + \frac{1}{\alpha R} \right) e^{-2\alpha R} + \frac{1}{\alpha^2 R} \left(1 + \frac{1}{\alpha} \right) + \frac{1}{\alpha} \left(1 - \frac{1}{\alpha} \right) \right] \\
&= \frac{N_0}{2} \left\{ \xi(\xi - 1) + (1 + \xi) \left[\frac{1}{R} - \left(\frac{1}{R} + \xi \right) e^{-2\xi R} \right] \right\}.
\end{aligned} \tag{167}
$$

- The self-energy integral Eq. (146) unfolds as

$$
\begin{aligned}
U &= \left\langle \psi_A \middle| \frac{1}{r_A} \middle| \psi_A \right\rangle = \frac{\gamma}{\pi} \int \frac{e^{-2\alpha r_A}}{r_A} \mathrm{d}\tau \\
&= \frac{\gamma R^3}{8\pi} \int_1^{\infty} \int_{-1}^{+1} \int_0^{2\pi} e^{-\alpha R(\lambda+\delta)} \frac{2}{R(\lambda+\delta)} (\lambda^2 - \delta^2) \mathrm{d}\lambda \mathrm{d}\delta \mathrm{d}\varphi \\
&= \frac{\gamma R^2}{2} \int_1^{\infty} \int_{-1}^{+1} e^{-\alpha R(\lambda+\delta)} (\lambda - \delta) \mathrm{d}\lambda \mathrm{d}\delta \\
&= \frac{\gamma R^2}{2} \left[\left(\int_1^{\infty} \lambda e^{-\alpha R \lambda} \mathrm{d}\lambda \right) \left(\int_{-1}^{+1} e^{-\alpha R \delta} \mathrm{d}\delta \right) - \left(\int_1^{\infty} e^{-\alpha R \lambda} \mathrm{d}\lambda \right) \left(\int_{-1}^{+1} \delta e^{-\alpha R \delta} \mathrm{d}\delta \right) \right] \\
&= \frac{\gamma}{2\alpha} \left[\left(\frac{1}{\alpha} - 1 \right) \left(1 + \frac{1}{\alpha R} \right) e^{-2\alpha R} + \frac{1}{\alpha R} \left(1 - \frac{1}{\alpha} \right) + \left(1 + \frac{1}{\alpha} \right) \right] \\
&= \frac{N_0}{2} \left\{ \xi(\xi + 1) + (\xi - 1) \left[\frac{1}{R} - \left(\frac{1}{R} + \xi \right) e^{-2\xi R} \right] \right\}.
\end{aligned} \tag{168}
$$

Before moving on to numerical evaluation, the bonding energetic integrals in the short- and long-long range interaction ($R \to \infty$) can still be evaluated analytically. Thus, by inspection of Eqs. (165)–(168), one finds the limits:

$$
S \to \begin{cases} 0 \ldots R \to \infty \\ N_0 \ldots R \to 0, \end{cases} \tag{169}
$$

$$
J \to \begin{cases} 0 \ldots R \to \infty \\ \frac{N_0}{a_0} \xi^2 \ldots R \to 0, \end{cases} \tag{170}
$$

$$
I \to \begin{cases} \frac{N_0}{a_0} \frac{\xi(\xi-1)}{2} \ldots R \to \infty \\ -\frac{N_0}{a_0} \xi \ldots R \to 0, \end{cases} \tag{171}
$$

$$U \to \begin{cases} \frac{N_0}{a_0}\frac{\xi(\xi+1)}{2} \ldots R \to \infty \\ \frac{N_0}{a_0}\xi \ldots R \to 0. \end{cases} \quad (172)$$

Under these conditions, by substituting Eqs. (169)–(172) back into the basic *physical and chemical bonding condensate formulations* Eqs. (155) and (156), one notes the equalization of the bonding and antibonding BEC levels under the long-range conditions, respectively, as follows:

$$\begin{aligned} E_{\text{bond-BEC-I}}^{\pm(R\to\infty)} &= -|\langle\psi(\mathbf{r})\rangle|^2 \frac{\eta_{\text{Molec}}}{8\pi} \frac{-2U - 2I \mp 4J}{E_0(1 \pm S) - U - 2I \mp 3J} \\ &= |\langle\psi(\mathbf{r})\rangle|^2 \frac{\eta_{\text{Molec}}}{2\pi} \frac{\xi^2 N_0}{2a_0E_0 + \xi N_0 - 3\xi^2 N_0}. \end{aligned} \quad (173)$$

$$\begin{aligned} E_{\text{bond-BEC-II}}^{\pm(R\to\infty)} &= E_0 - \frac{U + 2I \pm 3J}{1 \pm S} - E_{\text{bond-BEC-I}}^{\pm(R\to\infty)} \\ &= E_0 + \frac{N_0\xi(1-3\xi)}{2a_0} - E_{\text{bond-BEC-I}}^{\pm(\infty)}. \end{aligned} \quad (174)$$

However, for the short-range interaction, one obtains a *physical condensate* with equal bonding–antibonding energies:

$$E_{\text{bond-BEC-I}}^{\pm(R\to 0)} = -|\langle\psi(\mathbf{r})\rangle|^2 \frac{\eta_{\text{Molec}}}{8\pi}, \quad (175)$$

while the *chemical condensate appears forbidden* due to the infinite limits for the bonding and antibonding cases:

$$E_{\text{bond-BEC-II}}^{\pm(R\to 0)} \to \infty. \quad (176)$$

Now, to evaluate the BEC-bonding potential curves for bonding and antibonding energies as provided by the present model, one must recall the repulsive–attractive potential in a diatomic molecule of the form:

$$V(r) = \frac{A}{r^n} - \frac{B}{r^m}, \quad n > m. \quad (177)$$

For the equilibrium binding distance R and the energy E_{bond}, one has the system:

$$\begin{cases} \left(\frac{\partial V(r)}{\partial r}\right)_{r=R} = 0 \\ V(r = R) = E_{\text{bond}}, \end{cases} \quad (178)$$

with the solution:

$$A = \frac{m}{m-n} E_{\text{bond}} R^n, \ B = \frac{n}{m-n} E_{\text{bond}} R^m, \tag{179}$$

leading to the so-called Mie equation [92]

$$V(r) = \frac{E_{\text{bond}}}{m-n} \left[m \left(\frac{R}{r} \right)^n - n \left(\frac{R}{r} \right)^m \right], \tag{180}$$

or specialized to the celebrated Lennard–Jones 12–6 variant [93–99]

$$V^{12-6}(r) = -E_{bond} \left[\left(\frac{R}{r} \right)^{12} - 2 \left(\frac{R}{r} \right)^6 \right], \tag{181}$$

suitable for modeling H_2 and the inert gas interactions (He–He as well).

For practical implementation, one needs to set a few parameters, namely

- The bosonic order parameter may be, as a first approximation, set to unity, $|\langle \psi(\mathbf{r}) \rangle| = 1$, within the Tisza's phenomenological two-fluid model of Bose–Einstein condensation [100];
- The molecular chemical hardness may be set equal to that of atomic species entering homopolar binding, according to molecule softness (s) based formulation [101] rooted in the Bratsch atoms-in-molecule approach [102]:

$$\eta_{\text{AIM}} = \left. \frac{\sum_A s_A \eta_A}{\sum_A s_A} \right|_{\text{all } A \text{ equal}} = \eta_A. \tag{182}$$

- The equilibrium length R and the atomic ground-state energy E_0 are obtained from experimental or accurate quantum mechanical computations;
- The bonding energy E_{bond} is determined by physical and/or chemical BEC formulations for diatomic systems with one or two electrons in the valence shell; hydrogen and helium are the main prototypes.

Specifically, for atomic systems such as H and He as well as H–H and He–He bindings, the working and calculated values are summarized in Table 4. The results presented in Table 4 reveal interesting features about the physical and chemical bonding BEC at the level of hydrogen and helium systems, such as

- The inversion of the bonding (+) and antibonding (−) states, in the sense that the latter appears more "deeply" energetic than the former, is a sign of the nonbonding of the respective states; this is the case for the *physical bonding BEC(I)* of H_2 as well as for both the physical and chemical bonding BECs (I and II) of He_2, thus confirming that in this sophisticated way the He–He interaction is not

Table 4 Atomic and molecular data (number of electrons in atomic valence shell, atomic chemical hardness η, atomic ground-state energy E_0, interatomic homopolar binding length with respect to the first Bohr radius a_0, along the specific physical and chemical bonding BEC energies) for hydrogen and helium toward assessing their BEC-bonding Lennard–Jones-type potential (181), all in atomic units

Atoms	N_0	η[a]	E_0	*Molecules*	R	E_{bond}	$E^{+}_{BEC\text{-}I}$	$E^{-}_{BEC\text{-}I}$	$E^{+}_{BEC\text{-}II}$	$E^{-}_{BEC\text{-}II}$
H	1	0.237	−0.5	**H–H**	1.3[c]	−1.324[c]	−0.011	−0.017	−2.054	−1.513
He	2	0.458	−2.9[b]	**He–He**	5.67	-2.85×10^{-5}[d]	-8.15×10^{-3}	-1.3×10^{-2}	−3.695	−4.072

[a]Computed using finite-difference formula 0.5(I-A) for ionization and affinity (I and A, respectively) energies, spectroscopically determined in electron volts and then converted in atomic units

[b]From [103–109]

[c]From [110]

[d]Converted into Hartree from the graphical values shown in Fig. 2, with the conversion factor $x = k_B/E_h = 3.1668114(29) \times 10^{-6}$

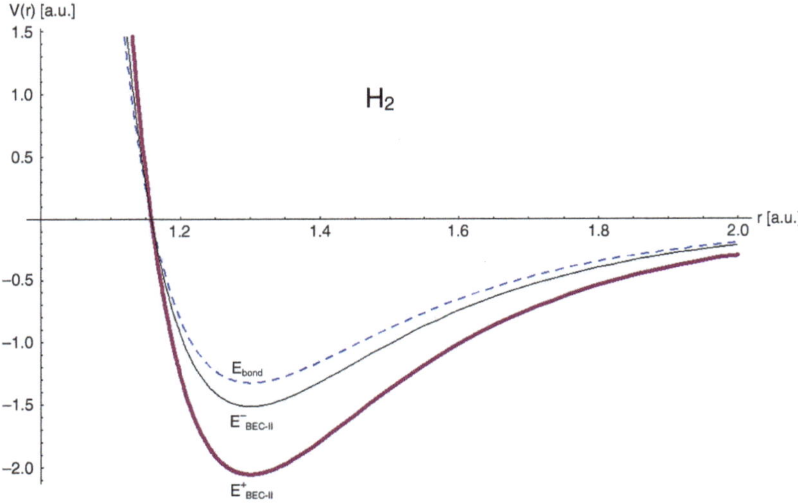

Fig. 5 Molecular hydrogen BEC bonding (*thick line*) and antibonding (*thin line*) potentials compared with ordinary or "normal" chemical bonding potentials (*dashed line*) as plotted from Eq. (181) using data presented in Table 4, respectively

possible in the sense of interatomic bonding, either in the physical or chemical sense

- However, the present BEC approach provides correction to chemical bonding of molecular hydrogen (BEC-II), showcasing the correct bonding vs. antibonding hierarchy; nevertheless, it reveals the interesting feature of having both of these states below the "normal" H_2 bonding. This is a sign that the BEC of chemical bonding provides an even more stable system and should be further studied at both theoretical and experimental levels to ultimately reconcile the associated 12–6 potentials abstracted from Eq. (181) with the data presented in Table 4, see Fig. 5.

5 Conclusions

Due to the increasing interest in low-temperature atomic and molecular condensates, an associated density-based theory is needed to systematically model and predict the bosonic condensate behavior of various interaction-trapping combinations. In this context, the finite-temperature Density Functional Theory of Bose Einstein Condensates (DFT–BEC) was here advanced by considering BEC specific order parameter and density concepts as variables in the celebrated Kohn–Sham (KS) equations due to their conceptual robustness with respect to both BEC and KS systems; however, because the KS system departs from the Hohenberg–Kohn (HK) system (with interparticle interaction), the HK–KS difference gives rise to the exchange-correlation (XC) density functional, which turns out to be an essential ingredient of the working KS equations and can be assessed by finite-temperature perturbation theory. For practical dilute bosonic systems, the XC functional is restricted to the exchange (X) functional as the first-order perturbation expansion for the HK–KS difference. The X-functional of BEC is then specialized within the so-called local density approximation (LDA) because it locally depends on the density and not on its gradient, being in this respect related to the Thomas–Fermi (TF) approximation as a limiting case of DFT. However, for the DFT–BEC combined approach, the TF limit is reached by neglecting in the KS-BEC equations both the density gradient and the bosonic interactions in the condensate.

The present picture may be further expanded physically in various ways. For instance, the homogeneous LDA version is considered to feature the working analytical exchange and current expressions for the KS-BEC equations, while providing the DFT–BEC benchmark to be further employed through variational perturbation theory for analytical traps [111–118]; further inclusion of the anomalous term in KS current and potential should account for strongly interacting bosons and allows for further advancements. Equally, the present DFT–BEC algorithm is designed to be analytically solved and compared with alternative Hartree–Fock–Bogoliubov equations modeling dipolar bosonic interactions in analytical traps [119–123].

From the point of view of chemistry, the present DFT–BEC employs the recently introduced bondons—the quantum particles of the chemical bond—to reinterpret the molecular Hamiltonian as being composed of fermionic and bosonic condensation components within an orthogonal fermionic–bosonic/bondonic space (FB); this model produces a generalization of molecular orbital theory, here restricted to the homopolar interaction, based on the same context in which the (fermionic) density functional theory and the (bosonic) Bose–Einstein condensation are grounded, i.e., on the same quantum completing relationship of the systems' particle number. In this context, physically and chemically condensed chemical bond components were identified as generalizing the classical Heitler–London scheme of homopolar interaction, having implemented additional DFT and BEC information such as the chemical hardness and the order parameter, respectively; however, by comparing hydrogen with helium DFT–BEC bonding, it was revealed that in the latter case all bonding and antibonding levels were inversed (i.e., where antibonding states observed to occur at "deeper" energetic levels than bonding states), thus providing a novel, general, and effectively energetic criteria for characterizing an unstable or forbidden chemical bond to be formed (as is the He_2 case). The same situation applies to the physical bonding BEC of molecular hydrogen but not to its chemical bonding, where, nevertheless, the correct bonding–antibonding hierarchy was computed as being entirely situated below the classical energy level of bonding—thus predicting more stable levels for molecular hydrogen condensates. This confirmation is left for the experimental community to confirm, while, on the theoretical side, further bosonic–bondonic condensation treatments of many-atomic molecular systems by means of generalizing DFT–BEC-Hückel theory of chemical bonding is envisaged.

Still, even for the hydrogen molecular system, the difference between the physical and chemical bonding behavior with respect to the bondonic–bosonic condensate emphasizes the specificity of natural law in the chemical realm, which is not necessarily viewed as a reducible physical counterpart. This affirms once more that chemical structure and bonding are special manifestations of nature at the level of atoms and molecules or as the surrealist poet Lucian Blaga writes "an unveiled mystery more mysteries revealed."

Acknowledgements The author is grateful to Dr. Axel Pelster and Prof. Hagen Kleinert from the Free University of Berlin for stimulating discussions during the years 2010 and 2011 to clarify the role of DFT in BEC; equally, the German Service for Academic Exchanges (DAAD) and Romanian National Council for Scientific Research (CNCS-UEFISCDI) are kindly thanked for awarding the grants 322 A/05356/2011 and TE16/2010-2013 that supported the research for the present study at the Free University of Berlin and West University of Timisoara, respectively.

References

1. Bose SN (1924) Z Phys 26:178
2. Einstein A (1925) Sitzungsber K Preuss Akad Wiss 1:3
3. London F (1938) Phys Rev 54:947
4. Landau LD (1941) J Phys USSR 5:71
5. Bogoliubov N (1947) J Phys USSR 11:23
6. Feynman RP (1953) Phys Rev 91:1291
7. Penrose O, Onsager L (1956) Phys Rev 104:576
8. Lee TD, Huang K, Yang CN (1957) Phys Rev 106:1135
9. Pitaevsky LP (1961) Sov Phys JETP 13:451
10. Gross EP (1961) Nuovo Cimento 20:454
11. Kamerlingh-Onnes H (1911) Proc K Ned Akad Wet 13:1274
12. Kapitza P (1938) Nature (London) 141:74
13. Osheroff DD, Richardson RC, Lee DM (1972) Phys Rev Lett 28:885
14. Leggett AJ (1972) Phys Rev Lett 29:1227
15. Cohen-Tannoudji CN (1998) Rev Mod Phys 70:707
16. Anderson MH, Ensher JR, Matthews MR, Wieman CE, Cornell EA (1995) Science 269:198
17. Davis KB, Mewes M-O, Andrews MR, van Druten NJ, Durfee DS, Kurn DM, Ketterle W (1995) Phys Rev Lett 75:3969
18. Greiner M, Mandel O, Esslinger T, Hänsch TW, Bloch I (2002) Nature (London) 415:39
19. Pethick CJ, Smith H (2004) Bose-Einstein condensation in dilute gases. Cambridge University Press, Cambridge
20. Hohenberg P, Kohn W (1964) Phys Rev 136:B864
21. Kohn W, Sham LJ (1965) Phys Rev 140:A1133
22. Parr RG, Yang W (1989) Density-functional theory of atoms and molecules. Oxford University Press, New York
23. March NH (1992) Electron density theory of atoms and molecules. Academic, New York
24. Dreizler RM, Gross EKU (1990) Density functional theory. An approach to the quantum many-body problem. Springer, Heidelberg
25. Jones RO, Gunnarson O (1989) Rev Mod Phys 61:689
26. Hohenberg PC, Martin PC (1965) Ann Phys (New York) 34:291
27. Oliveira LN, Gross EKU, Kohn W (1988) Phys Rev Lett 60:2430

28. Jochim S, Bartenstein M, Altmeyer A, Hendl G, Riedl S, Chin C, Denschlag JH, Grimm R (2003) Science 302:2101
29. Greiner M, Regal CA, Jin DS (2003) Nature (London) 426:537
30. Regal CA, Greiner M, Jin DS (2004) Phys Rev Lett 92:040403
31. Bardeen J, Cooper LN, Schrieffer JR (1957) Phys Rev 106:162
32. Feshbach H (1964) Rev Mod Phys 36:1076
33. Grimme S (2004) J Comput Chem 25:1463
34. Putz MV (2008) Int J Mol Sci 9:1050
35. Vetter A (1997) Density Functional Theory for BEC (in German). Diploma Thesis, Institute of Theoretical Physics, Bayerische Julius-Maximilians University, Würzburg
36. Nunes GS (1999) J Phys B: At Mol Opt Phys 32:4293
37. Albus AP, Illuminati F, Wilkens M (2003) Phys Rev A 67:063606
38. Kim YE, Zubarev AL (2003) Phys Rev A 67:015602
39. Brand J (2004) J Phys B: At Mol Opt Phys 37:S287
40. Argaman N, Band YB (2011) Phys Rev A 83:023612
41. Putz MV (2012) In: Chemical Information and Computational Challenges. Putz MV (Ed), NOVA Science, New York
42. Putz M.V. (2011) In: Advances in chemistry research. Vol. 10. Taylor JC (Ed), NOVA Science, New York
43. Putz MV (2012) Int J Chem Model 4(1):000–000
44. Putz MV (2012) Quantum theory: density, condensation, and bonding. Apple Academics & CRC Press—Taylor & Francis Group, Toronto & New Jersey
45. Huang K (2001) Introduction to statistical physics. Taylor and Francis, London
46. Thomas LH (1927) Proc Cambridge Phil Soc 23:542
47. Fermi E (1927) Rend Accad Naz Lincei 6:602
48. Anderson P (1972) Sci New Ser 177:393
49. Kadanoff LP (2009) J Stat Phys 137:777
50. Park JH, Kim SW (2010) Phys Rev A 81:063636
51. Chermette H (1999) J Comput Chem 20:129
52. Geerlings P, De Proft F, Langenaeker W (2003) Chem Rev 103:1793
53. Putz MV (2011) MATCH Commun Math Comput Chem 66:35
54. Parr RG, Pearson RG (1983) J Am Chem Soc 105:7512
55. Berkowitz M, Ghosh SK, Parr RG (1985) J Am Chem Soc 107:6811
56. Baekelandt BG, Cedillo A, Parr RG (1995) J Chem Phys 103:8548
57. Pauling L (1931) J Am Chem Soc 53:3225
58. Gillespie RJ (1970) J Chem Educ 47:18
59. Berlin T (1951) J Chem Phys 19:208
60. Daudel R (1980) In: Electron and magnetization densities in molecules and crystals. Becker P (Ed) NATO ASI, Series B-Physics, Volume 40, Plenum Press, New York
61. Bader RFW (1990) Atoms in molecules: a quantum theory, vol 22, 1st edn. The international series of monographs on chemistry, Clarendon, Oxford
62. Putz MV (2008) Symmetry: Cult Sci 19:249
63. Putz MV (2010) Int J Mol Sci 11:4227
64. Putz MV (2012) In: Nanoscience and advancing computational methods in chemistry. Castro EA, Haghi AK (Eds), IGI Global, Pasadena
65. de Broglie L (1923) Compt Rend Acad Sci (Paris) 177:507
66. de Broglie L (1925) Compt Rend Acad Sci (Paris) 180:498
67. de Broglie L, Vigier MJP (1953) La Physique Quantique Restera-t-elle Indéterministe? Gauthier-Villars, Paris
68. Bohm D, Vigier JP (1954) Phys Rev 96:208
69. Putz MV (2011) In: Quantum frontiers of atoms and molecules. Putz MV (Ed), NOVA Science, New York, pp. 1–20

70. Oelke WC (1969) Laboratory physical chemistry. Van Nostrand Reinhold Company, New York
71. Findlay A (1955) Practical physical chemistry. Longmans, London
72. Putz MV (2010) Int J Mol Sci 11:4124
73. Whitney CK (2007) Found Phys 37:788
74. Lennard-Jones JE (1929) Trans Faraday Soc 25:668
75. Hartree DR (1957) The calculation of atomic structures. Wiley, New York
76. Slater JC (1963) Theory of molecules and solids. Electronic structure of molecules, vol 1. McGraw-Hill, New York
77. Roothaan CCJ (1960) Rev Mod Phys 32:179
78. Pople JA, Nesbet RK (1954) J Chem Phys 22:571
79. Corongiu G (2007) J Phys Chem A 111:5333
80. Glaesemann KR, Schmidt MW (2010) J Phys Chem A 114:8772
81. Slater JC, Kirkwood JG (1931) Phys Rev 37:682
82. Modugno G, Ferrari G, Roati G, Brecha RJ, Simoni A, Inguscio M (2001) Science 294:1320
83. Heitler W, London F (1927) Z Phys 44:455
84. Ruedenberg K (1962) Rev Mod Phys 34:326
85. Hugenholtz NM, Pines D (1959) Phys Rev 116:489
86. Leggett AJ (2001) Rev Mod Phys 73:307
87. Chattaraj PK, Lee H, Parr RG (1991) J Am Chem Soc 113:1854
88. Mulliken RS, Riecke CA, Orloff D, Orloff H (1949) J Chem Phys 17:1248
89. Roberts JL, Jaffe H (1957) J Chem Phys 27:883
90. Parr RG (1972) The quantum theory of molecular electronic structure. WA Benjamin, Reading, MA
91. Putz MV (2003) Contributions within density functional theory with applications to chemical reactivity theory and electronegativity. Dissertation, Parkland
92. Borg RJ, Diens GJ (1992) The physical chemistry of solids. Academic, Boston
93. Lennard-Jones JE (1924) Proc R Soc Lond A 106(738):463
94. Lennard-Jones JE (1931) Proc Phys Soc 43(5):461
95. Barron THK, Domb C (1955) Proc R Soc Lond A Math Phys Sci 227(1171):447
96. Smit B (1992) J Chem Phys 96:8639
97. Atkins P, de Paula J (2006) Atkins' physical chemistry, 8th edn. Oxford University Press, Oxford
98. Frenkel D, Smit B (2002) Understanding molecular simulation, 2nd edn. Academic, San Diego
99. Zhen S, Davies GJ (1983) Phys Status Solidi A 78:595
100. Tisza L (1938) Nature (London) 141:913
101. Putz MV (2011) Curr Phys Chem 1:111
102. Bratsch SG (1985) J Chem Educ 62:101
103. Bransden BH, Joachain CJ (2003) Physics of atoms and molecules, 2nd edn. Pearson Education, Essex
104. Drake GWF, Van Z-C (1994) Chem Phys Lett 229:486
105. Yan Z-C, Drake GWF (1995) Phys Rev Lett 74:4791
106. Drake GWF (1999) Phys Scr T83:83
107. Baker JD, Freund DE, Hill RN, Morgan JD III (1990) Phys Rev A 41:1247
108. Scott TC, Lüchow A, Bressanini D, Morgan JD III (2007) Phys Rev A 75:060101
109. Drake GWF (2006) Springer handbook of atomic, molecular, and optical physics. Drake GWF (Ed). Springer, New York
110. Condon EU (1927) Proc Natl Acad Sci 13:466
111. Stevenson PM (1981) Phys Rev D 23:2916
112. Feynman RP, Kleinert H (1986) Phys Rev A 34:5080
113. Kleinert H, Pelster A, Putz MV (2002) Phys Rev E 65:066128

114. Kleinert H (2008) Path integrals in quantum mechanics, statistics, polymer physics, and financial markets, 5th edn. World Scientific, Singapore, Chapter 5
115. Kleinert H, Schulte-Frohlinde V (2001) Critical properties of φ^4-theories. World Scientific, Singapore
116. Hamprecht B, Pelster A (2001) In: Fluctuating Paths and Fields. Janke W, Pelster, A, Schmidt H-J, Bachmann M (Eds), World Scientific, Singapore, p. 347
117. Kleinert H (2003) Mod Phys Lett B 17:1011
118. Putz MV (2009) Int J Mol Sci 10:4816
119. Bagnato V, Pritchard DE, Kleppner D (1987) Phys Rev A 35:4354
120. Stringari S (1996) Phys Rev Lett 77:2360
121. Yi S, You L (2001) Phys Rev A 63:053607
122. Andersen JO (2004) Rev Mod Phys 76:599
123. Lima ARP, Pelster A (2011) Phys Rev A 84:041604(R)

Struct Bond (2012) 149: 51–94
DOI: 10.1007/978-3-642-32753-7_2

Information Theory Insights into Molecular Electronic Structure and Reactivity

Roman F. Nalewajski

Abstract Selected concepts and techniques of *Information-Theory* (IT) are summarized and their use in probing the molecular electronic structure is advocated. The electron redistributions accompanying formation of chemical bonds, relative to the (molecularly placed) free atoms of the corresponding "promolecule," generate the associated displacements in alternative measures of the amount of information carried by electrons. The latter are shown to provide sensitive probes of information origins of the chemical bonds, allow the spatial localization of bonding regions in molecules, and generate attractive entropy/information descriptors of the system bond multiplicities. Information-theoretic descriptors of both the molecule as a whole and its diatomic fragments can be extracted. Displacements in the molecular Shannon entropy and entropy deficiency, relative to the promolecular reference, are investigated. Their densities provide efficient tools for detecting the presence of the direct chemical bonds and for monitoring the promotion/hybridization changes the bonded atoms undergo in a molecular environment. The nonadditive Fisher information density in the *Atomic Orbital* (AO) resolution is shown to generate an efficient *Contra-Gradience* (CG) probe for locating the bonding regions in molecules. Rudiments of the *Orbital Communication Theory* (OCT) of the chemical bond are introduced. In this approach molecules are treated as information systems propagating "signals" of electron allocations to basis functions, from AO "inputs" to AO "outputs." The conditional probabilities defining such an information network are generated using the bond-projected superposition principle of quantum mechanics. They are proportional to squares of the corresponding elements of the first-order density matrix in AO representation. Therefore, they are related to Wiberg's quadratic index of the chemical bond multiplicity. Such information propagation in molecules exhibits typical communication "noise" due to the electron delocalization via the system

R.F. Nalewajski (✉)
Department of Theoretical Chemistry, Jagiellonian University, R. Ingardena 3, 30-060 Cracow, Poland
e-mail: nalewajs@chemia.uj.edu.pl

chemical bonds. In describing this scattering of electron probabilities throughout the network of chemical bonds, due to the system occupied *Molecular Orbitals* (MO), the OCT uses the standard entropy/information descriptors of communication devices. They include the average communication noise (IT covalency) and information flow (IT ionicity) quantities, reflected by the channel conditional entropy and mutual information characteristics, respectively. Recent examples of applying these novel tools in an exploration of the electronic structure and bonding patterns of representative molecules are summarized. This communication perspective also predicts the *"indirect"* (through-*bridge*) sources of chemical interactions, due to the *"cascade"* probability propagation realized via AO intermediates. It supplements the familiar through-*space* mechanism, due to the constructive interference between the interacting AO, which generates the *"direct"* communications between bonded atoms. Such bridge "bonds" effectively extend the range of chemical interactions in molecular systems. Representative examples of the π systems in benzene and butadiene are discussed in a more detail and recent applications of the information concepts in exploring the elementary reaction mechanisms are mentioned.

Keywords Bond information probes • Bond localization • Chemical bonds • Chemical reactivity • Contra-gradience criterion • Covalent/ionic bond components • Direct/indirect bond multiplicities • Entropic bond indices • Fisher information • Information theory • Molecular information channels • Orbital communications

Contents

1 Introduction

The Information Theory (IT) [1–8] is one of the youngest branches of the applied probability theory, in which the probability ideas have been introduced into the field of communication, control, and data processing. Its foundations have been laid in 1920s by Sir R. A. Fisher [1] in his classical measurement theory and in 1940s by C.E. Shannon [3] in his mathematical theory of communication. The quantum state of electrons in a molecule is determined by the system wave function, the amplitude of the particle probability distribution which carries the information. It is thus intriguing to explore the information content of electronic probability distributions in molecules and to extract from it the pattern of chemical bonds, reactivity trends and other molecular descriptors, e.g., the bond multiplicities ("orders") and their covalent/ionic composition. In this survey we summarize recent applications of IT in probing chemical bonds in molecules. In particular, changes in the information content due to subtle electron redistributions accompanying the bond formation process will be examined. Elsewhere, e.g., [9–14], it has been amply demonstrated that many classical problems of theoretical chemistry can be approached afresh using such a novel IT perspective. For example, the displacements in the information distribution in molecules, relative to the *promolecular* reference consisting of the nonbonded constituent atoms, have been investigated [9–16] and the least-biased partition of the molecular electron distributions into subsystem contributions, e.g., densities of bonded Atoms-in-Molecules (AIM), have been investigated [9, 17–24]. These optimum density pieces have been derived from alternative global and local variational principles of IT. The IT approach has been shown to lead to the "stockholder" molecular fragments of Hirshfeld [25].

The spatial localization of specific bonds presents another challenging problem to be tackled by this novel treatment of molecular systems. Another diagnostic problem in the theory of molecular electronic structure deals with the shell structure and electron localization in atoms and molecules. The nonadditive Fisher information in the Atomic Orbital (AO) resolution has been recently used as the Contra-Gradience (CG) criterion for localizing the bonding regions in molecules [10–14, 26–28], while the related information density in the Molecular Orbital (MO) resolution has been shown [9, 29] to determine the vital ingredient of the Electron-Localization Function (ELF) [30–32].

The Communication Theory of the Chemical Bond (CTCB) has been developed using the basic entropy/information descriptors of molecular information (communication) channels in the AIM, orbital and local resolutions of the electron probability distributions [9–11, 33–48]. The same bond descriptors have been used to provide the information-scattering perspective on the intermediate stages in the electron redistribution processes [49], including the atom promotion via the orbital hybridization [50], and the communication theory for the excited electron configurations has been developed [51]. Moreover, a phenomenological treatment of equilibria in molecular subsystems has been proposed [9, 52–54], which formally resembles the ordinary thermodynamic description.

Entropic probes of the molecular electronic structure have provided attractive tools for describing the chemical bond phenomenon in information terms. It is the main purpose of this survey to summarize alternative local entropy/information probes of the molecular electronic structure [9–14, 21, 22] and to explore the information origins of the chemical bonds. It is also our goal to present recent developments in the AO formulation of CTCB, called the Orbital Communication Theory (OCT) [10, 11, 39, 48, 55–58]. The importance of the nonadditive effects in the chemical-bond phenomena will be emphasized and the information-cascade (bridge) propagation of electronic probabilities in molecular information systems, generating the indirect bond contributions due to the orbital intermediaries [59–63], will be examined. Throughout the article symbol A denotes a scalar quantity, $\boldsymbol{A}$ stands for the *row*-vector, and $\mathbf{A}$ represents a square or rectangular matrix. In the logarithmic measure of information the logarithm is taken to base 2, $\log = \log_2$, which expresses the amount of information in *bits* (binary digits). Accordingly, selecting $\log = \ln$ measures the information in *nats* (natural units): 1 nat = 1.44 bits.

2 Measures of Information Content

We begin with a short summary of selected IT concepts and techniques to be used in diagnosing the information content of electronic distributions in molecules and in probing their chemical bonds. The *Shannon entropy* [3, 4] in the (normalized) discrete probability vector $\boldsymbol{p} = \{p_i\}$,

$$S(\boldsymbol{p}) = -\sum\nolimits_i p_i \log p_i, \qquad \sum\nolimits_i p_i = 1, \tag{1}$$

where the summation extends over labels of all elementary events determining the probability scheme in question, provides a measure of the average indeterminacy in $\boldsymbol{p}$. This function also measures the average amount of information obtained when the uncertainty is removed by an appropriate measurement (experiment).

The Fisher information for locality [1, 2], called the *intrinsic accuracy*, historically predates the Shannon entropy by about 25 years, being proposed in about the same time when the final form of quantum mechanics was shaped. It emerges as an expected error in a "smart" measurement, in the context of an efficient estimator of a parameter. For the "locality" parameter, the Fisher measure of the information content in the continuous (normalized) probability density $p(\boldsymbol{r})$ reads:

$$I[p] = \int p(\boldsymbol{r})[\nabla \ln p(\boldsymbol{r})]^2 \mathrm{d}\boldsymbol{r} = \int [\nabla p(\boldsymbol{r})]^2/p(\boldsymbol{r})\mathrm{d}\boldsymbol{r}, \qquad \int p(\boldsymbol{r})\,\mathrm{d}\boldsymbol{r} = 1. \tag{2}$$

This Fisher functional can be simplified by expressing it as the functional of the associated classical (real) amplitude $A(\boldsymbol{r}) = [p(\boldsymbol{r})]^{-1/2}$ of the probability distribution:

$$I[p] = 4\int [\nabla A(\boldsymbol{r})]^2 \mathrm{d}\boldsymbol{r} \equiv I[A]. \tag{3}$$

The amplitude form is then naturally generalized into the domain of complex probability amplitudes, e.g., the wave functions encountered in quantum mechanics [2, 26]. For the simplest case of the spinless *one*-particle system, when $p(\boldsymbol{r}) = \psi^*(\boldsymbol{r})\psi(\boldsymbol{r}) = |\psi(\boldsymbol{r})|^2$,

$$I[\psi] = 4\int |\nabla\psi(\boldsymbol{r})|^2 \mathrm{d}\boldsymbol{r} = 4\int \nabla\psi^*(\boldsymbol{r}) \cdot \nabla\psi(\boldsymbol{r})\mathrm{d}\boldsymbol{r} \equiv \int f(\boldsymbol{r})\mathrm{d}\boldsymbol{r}. \tag{4}$$

The Fisher information is reminiscent of von Weizsäcker's [64] inhomogeneity correction to the electronic kinetic energy in the Thomas–Fermi theory. It characterizes the compactness of the probability density $p(\boldsymbol{r})$. For example, the Fisher information in the familiar *normal distribution* measures the inverse of its variance, called the *invariance*, while the complementary Shannon entropy is proportional to the logarithm of variance, thus monotonically increasing with the spread of the Gaussian distribution. Therefore, the Shannon entropy and intrinsic accuracy describe complementary facets of the probability density: the former reflects the distribution "spread," providing a measure of its uncertainty ("disorder"), while the latter measures a "narrowness" ("order") in the probability density.

An important generalization of Shannon's entropy, called the relative (*cross*) entropy, also known as the *entropy deficiency*, *missing information* or the *directed divergence*, has been proposed by Kullback and Leibler [5, 6]. It measures the information "distance" between the two (normalized) probability distributions for the same set of events. For example, in the discrete probability scheme identified by events $\boldsymbol{a} = \{a_i\}$ and their probabilities $\boldsymbol{P}(\boldsymbol{a}) = \{P(a_i) = p_i\} \equiv \boldsymbol{p}$, this discrimination information in $\boldsymbol{p}$ with respect to the reference distribution $\boldsymbol{p}^0 = \{P^0(a_i) = p_i^{\,0}\}$ reads:

$$\Delta S(\boldsymbol{p}|\boldsymbol{p}^0) = \sum\nolimits_i p_i \log\left(p_i/p_i^{\,0}\right) \geq 0. \tag{5}$$

The entropy deficiency provides a measure of the information resemblance between the two compared probability schemes. The more the two distributions differ from one another, the larger the information distance. For individual events the logarithm of the probability ratio $I_i = \log(p_i/p_i^{\,0})$, called the *surprisal*, provides a measure of the event information in the current distribution relative to that in the reference distribution. The equality in the preceding equation takes place only for the vanishing surprisal for all events, i.e., when the two probability distributions are identical.

For two mutually dependent (discrete) *probability vectors* $\boldsymbol{P}(\boldsymbol{a}) = \{P(a_i) = p_i\} \equiv \boldsymbol{p}$ and $\boldsymbol{P}(\boldsymbol{b}) = \{P(b_j) = q_j\} \equiv \boldsymbol{q}$, one decomposes the joint probabilities of the simultaneous events $\boldsymbol{a}\wedge\boldsymbol{b} = \{a_i\wedge b_j\}$ in the two schemes, $\mathbf{P}(\boldsymbol{a}\wedge\boldsymbol{b}) = \{P(a_i\wedge b_j) = \pi_{i,j}\} \equiv \boldsymbol{\pi}$, as products of the "marginal" probabilities of events in one set, say $\boldsymbol{a}$, and the corresponding conditional probabilities $\mathbf{P}(\boldsymbol{b}|\boldsymbol{a}) = \{P(j|i)\}$ of outcomes in another set $\boldsymbol{b}$, given that events $\boldsymbol{a}$ have already occurred: $\{\pi_{i,j} = p_i\ P(j|i)\}$. Relevant normalization conditions for the joint and conditional probabilities then read:

$$\begin{aligned} &\sum\nolimits_j \pi_{i,j} = p_i, \quad \sum\nolimits_i \pi_{i,j} = q_j, \quad \sum\nolimits_i \sum\nolimits_j \pi_{i,j} = 1, \\ &\sum\nolimits_j P(j|i) = 1, \quad i = 1, 2, \ldots \end{aligned} \tag{6}$$

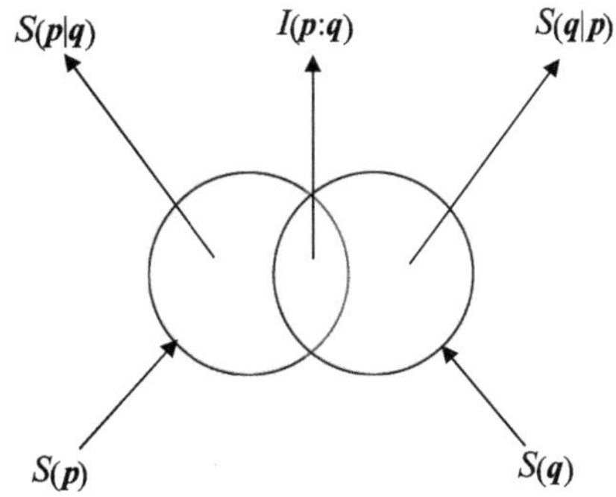

Fig. 1 Diagram of the conditional entropy and mutual information quantities for two dependent probability distributions $\boldsymbol{p}$ and $\boldsymbol{q}$. *Two circles* enclose areas representing the entropies $S(\boldsymbol{p})$ and $S(\boldsymbol{q})$ of the separate probability vectors, while their common (overlap) area corresponds to the mutual information $I(\boldsymbol{p}{:}\boldsymbol{q})$ in two distributions. The remaining part of each circle represents the corresponding conditional entropy, $S(\boldsymbol{p}|\boldsymbol{q})$ or $S(\boldsymbol{q}|\boldsymbol{p})$, measuring the residual uncertainty about events in one set, when one has the full knowledge of the occurrence of events in the other set of outcomes. The area enclosed by the circle envelope then represents the entropy of the "product" (joint) distribution: $S(\boldsymbol{\pi}) = S(\mathbf{P}(\boldsymbol{a} \wedge \boldsymbol{b})) = S(\boldsymbol{p}) + S(\boldsymbol{q}) - I(\boldsymbol{p} : \boldsymbol{q}) = S(\boldsymbol{p}) + S(\boldsymbol{q}|\boldsymbol{p}) = S(\boldsymbol{q}) + S(\boldsymbol{p}|\boldsymbol{q})$.

The Shannon entropy of this "product" distribution,

$$\begin{aligned} S(\boldsymbol{\pi}) &= -\sum_i \sum_j \pi_{i,j} \log \pi_{i,j} = -\sum_i \sum_j p_i P(j|i)[\log p_i + \log P(j|i)] \\ &= -\sum_j P(j|i)] \sum_i p_i \log p_i - \sum_i p_i \left[\sum_j P(j|i) \log P(j|i)\right] \\ &= S(\boldsymbol{p}) + \sum_i p_i S(\boldsymbol{q}|i) \equiv S(\boldsymbol{p}) + S(\boldsymbol{q}|\boldsymbol{p}), \end{aligned} \tag{7}$$

has been expressed above as the sum of the average entropy $S(\boldsymbol{p})$ in the marginal probability distribution, and the average *conditional entropy* in $\boldsymbol{q}$ given $\boldsymbol{p}$:

$$S(\boldsymbol{q}|\boldsymbol{p}) = -\sum_i \pi_{i,j} \log P(j|i)]. \tag{8}$$

The latter represents the extra amount of uncertainty about the occurrence of events $\boldsymbol{b}$, given that events $\boldsymbol{a}$ are known to have occurred. In other words, the amount of information obtained as a result of simultaneously observing events $\boldsymbol{a}$ and $\boldsymbol{b}$ of the two discrete probability distributions equals to the amount of information in one set, say $\boldsymbol{a}$, supplemented by the extra information provided by the occurrence of events in the other set $\boldsymbol{b}$, when $\boldsymbol{a}$ are known to have occurred already. This is qualitatively illustrated in Fig. 1 [7, 8].

The common amount of information in two events a_i and b_j, $I(i{:}j)$, measuring the information about a_i provided by the occurrence of b_j, or the information about b_j provided by the occurrence of a_i, is called the *mutual information* in two events:

$$I(i:j) = \log\left[P(a_i \wedge b_j)/P(a_i)P(b_j)\right] = \log\left[\pi_{i,j}/(p_i q_j)\right] \equiv \log[P(i|j)/p_i]$$
$$\equiv \log\left[P(j|i)/q_j\right] = I(j:i). \tag{9}$$

It may take on any real value, positive, negative, or zero. It vanishes when both events are independent, i.e., when the occurrence of one event does not influence (or condition) the probability of the occurrence of the other event, and it is negative when the occurrence of one event makes a nonoccurrence of the other event more likely. It also follows from the preceding equation that

$$I(i:j) = I(i) - I(i|j) = I(j) - I(j|i) = I(i) + I(j) - I(i \wedge j) \quad \text{or}$$
$$I(i \wedge j) = I(i) + I(j) - I(i:j), \tag{10}$$

where the *self-information* of the joint event $I(i \wedge j) = -\log\pi_{i,j}$. Hence, the information in the joint occurrence of two events a_i and b_j is the information in the occurrence of a_i plus that in the occurrence of b_j minus the mutual information. For independent events, when $\pi_{i,j} = p_i q_j$, $I(i{:}j) = 0$ and hence $I(i, j) = I(i) + I(j)$.

The mutual information of an event with itself defines its self-information: $I(i{:}i) \equiv I(i) = \log[P(i|i)/p_i] = -\log p_i$, since $P(i|i) = 1$. It vanishes when $p_i = 1$, i.e., when there is no uncertainty about the occurrence of a_i, so that the occurrence of this event removes no uncertainty and hence conveys no information. This quantity provides a measure of the uncertainty about the occurrence of the event itself, i.e., the information received when the event occurs. The Shannon entropy of Eq. (1) can be thus interpreted as the mean value of self-information in all individual events: $S(\boldsymbol{p}) = \sum_i p_i I(i)$. One similarly defines the average mutual information in two probability distributions as the $\boldsymbol{\pi}$-weighted mean value of the mutual information quantities for individual joint events:

$$I(\boldsymbol{p}:\boldsymbol{q}) = \sum_i \sum_j \pi_{i,j} \log(\pi_{i,j}/\pi_{i,j}{}^0) = S(\boldsymbol{p}) + S(\boldsymbol{q}) - S(\pi)$$
$$= S(\boldsymbol{p}) - S(\boldsymbol{p}|\boldsymbol{q}) = S(\boldsymbol{q}) - S(\boldsymbol{q}|\boldsymbol{p}) \geq 0, \tag{11}$$

where the equality holds only for the independent distributions, when $\pi_{i,j} = \pi_{i,j}{}^0 = p_i q_j$. Indeed, the amount of uncertainty in $\boldsymbol{q}$ can only decrease when $\boldsymbol{p}$ has been known beforehand, $S(\boldsymbol{q}) \geq S(\boldsymbol{q}|\boldsymbol{p}) = S(\boldsymbol{q}) - I(\boldsymbol{p}{:}\boldsymbol{q})$, with the equality being observed only when the two sets of events are independent (*non*overlapping). These average entropy/information relations are also illustrated in Fig. 1.

The average mutual information is an example of the entropy deficiency, measuring the missing information between the joint probabilities $\mathbf{P}(\boldsymbol{a} \wedge \boldsymbol{b}) = \boldsymbol{\pi}$ of the *dependent* events $\boldsymbol{a}$ and $\boldsymbol{b}$, and the joint probabilities $\mathbf{P}^{\text{ind}}(\boldsymbol{a} \wedge \boldsymbol{b}) = \boldsymbol{\pi}^0 = \boldsymbol{p}^{\text{T}}\boldsymbol{q}$ for the *independent* events: $I(\boldsymbol{p}{:}\boldsymbol{q}) = \Delta S(\boldsymbol{\pi}|\boldsymbol{\pi}^0)$. The average mutual information thus reflects a degree of dependence between events defining the compared probability schemes. A similar information-distance interpretation can be attributed to the conditional entropy: $S(\boldsymbol{p}|\boldsymbol{q}) = S(\boldsymbol{p}) - \Delta S(\boldsymbol{\pi}|\boldsymbol{\pi}^0)$.

3 Communication Systems

We continue this short introduction to IT with the entropy/information descriptors of a transmission of signals in communication systems [3, 4, 7, 8] (Fig. 2). The signal emitted from n "inputs" $\boldsymbol{a} = (a_1, a_2, \ldots, a_n)$ of the channel *source* **A** is characterized by the probability distribution $\boldsymbol{P}(\boldsymbol{a}) = \boldsymbol{p} = (p_1, p_2, \ldots, p_n)$. It can be received at m "outputs" $\boldsymbol{b} = (b_1, b_2, \ldots, b_m)$ of the system *receiver* **B**. A transmission of signals in the channel is randomly disturbed thus exhibiting the communication *noise*. Indeed, in general the signal sent at the given input can be received with a nonvanishing probability at several outputs. This feature of communication systems is described by the conditional probabilities of *outputs-given-inputs*, $\mathbf{P}(\boldsymbol{b}|\boldsymbol{a}) = \{P(b_j|a_i) = P(a_i \wedge b_j)/P(a_i) \equiv P(j|i)\}$, where $P(a_i \wedge b_j) \equiv \pi_{i,j}$ stands for the probability of the joint occurrence of the specified pair of the input–output events. The distribution of the output signal among the detection "events" $\boldsymbol{b}$ is thus given by the output probability distribution $\boldsymbol{P}(\boldsymbol{b}) = \boldsymbol{q} = (q_1, q_2, \ldots, q_m) = \boldsymbol{p}\,\mathbf{P}(\boldsymbol{b}|\boldsymbol{a})$.

The Shannon entropy $S(\boldsymbol{p})$ of the input (source) probabilities $\boldsymbol{p}$ determines the channel a priori entropy. The average *conditional entropy* $H(\mathbf{B}|\mathbf{A}) \equiv S(\boldsymbol{q}|\boldsymbol{p})$ of the outputs-given-inputs, determined by the scattering probabilities $\mathbf{P}(\boldsymbol{b}|\boldsymbol{a})$, then measures the average noise in the $\boldsymbol{a} \rightarrow \boldsymbol{b}$ transmission. The so-called *a posteriori* entropy $H(\mathbf{A}|\mathbf{B}) \equiv S(\boldsymbol{p}|\boldsymbol{q})$, of *inputs-given-outputs*, is similarly defined by the "reverse" conditional probabilities of the $\boldsymbol{b} \rightarrow \boldsymbol{a}$ transmission signals: $\mathbf{P}(\boldsymbol{a}|\boldsymbol{b}) = \{P(a_i|b_j) = P(i|j)\}$. It reflects the residual indeterminacy about the input signal, when the output signal has already been received. The average conditional entropy $S(\boldsymbol{p}|\boldsymbol{q})$ thus measures the indeterminacy of the source with respect to the receiver, while the conditional entropy $S(\boldsymbol{q}|\boldsymbol{p})$ reflects the uncertainty of the receiver relative to the source. Hence, an observation of the output signal provides on average the amount of information given by the difference between the a priori and *a posteriori* uncertainties, which defines the *mutual information* in the source and receiver: $S(\boldsymbol{p}) - S(\boldsymbol{p}|\boldsymbol{q}) = I(\boldsymbol{p}:\boldsymbol{q})$. This quantity measures the net amount of information transmitted through the communication channel, while the conditional information $S(\boldsymbol{p}|\boldsymbol{q})$ reflects a fraction of $S(\boldsymbol{p})$ transformed into "noise" as a result of the input signal being scattered in the channel. Accordingly, $S(\boldsymbol{q}|\boldsymbol{p})$ reflects the noise part of $S(\boldsymbol{q}) = S(\boldsymbol{q}|\boldsymbol{p}) + I(\boldsymbol{p}:\boldsymbol{q})$ (see Fig. 1).

Consider as an illustrative example the familiar Binary Channel of Fig. 3 defined by the symmetric conditional probability matrix of observing its two outputs, given the two inputs,

$$\mathbf{P}(\boldsymbol{b}|\boldsymbol{a}) = \begin{bmatrix} 1-\omega & \omega \\ \omega & 1-\omega \end{bmatrix}. \tag{12}$$

The given input probabilities $\boldsymbol{p} = (x, 1-x)$ characterize the way this communication device is used. Its input (a priori) entropy is determined by the *Binary Entropy Function* (BEF) $S(\boldsymbol{p}) = -x\log x - (1-x)\log(1-x) \equiv H(x)$ shown in Fig. 4. The

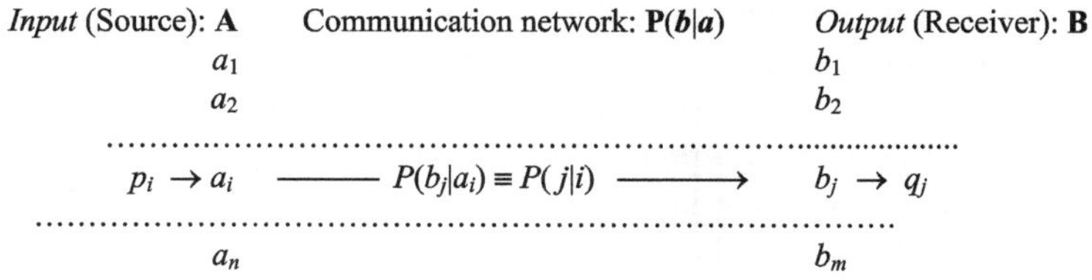

Fig. 2 Schematic diagram of the communication system characterized by two probability vectors: $\boldsymbol{P}(\boldsymbol{a}) = \{P(a_i)\} = \boldsymbol{p} = (p_1, \ldots, p_n)$, of the channel "input" events $\boldsymbol{a} = (a_1, \ldots, a_n)$ in the system source **A**, and $\boldsymbol{P}(\boldsymbol{b}) = \{P(b_j)\} = \boldsymbol{q} = (q_1, \ldots, q_m)$, of the "output" events $\boldsymbol{b} = (b_1, \ldots, b_m)$ in the system receiver **B**. The transmission of signals in this communication channel is described by the $(n \times m)$-matrix of the conditional probabilities $\mathbf{P}(\boldsymbol{b}|\boldsymbol{a}) = \{P(b_j|a_i) \equiv P(j|i)\}$, of observing different "outputs" (*columns*, $j = 1, 2, \ldots, m$), given the specified "inputs" (*rows*, $i = 1, 2, \ldots, n$). For clarity, only a single scattering $a_i \rightarrow b_j$ is shown in the diagram

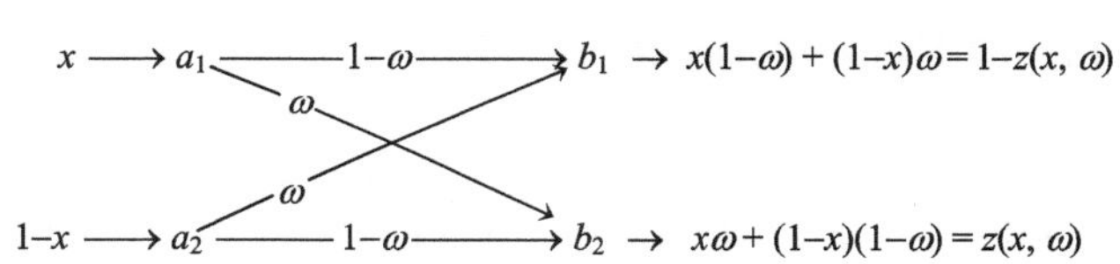

Fig. 3 The symmetric binary channel

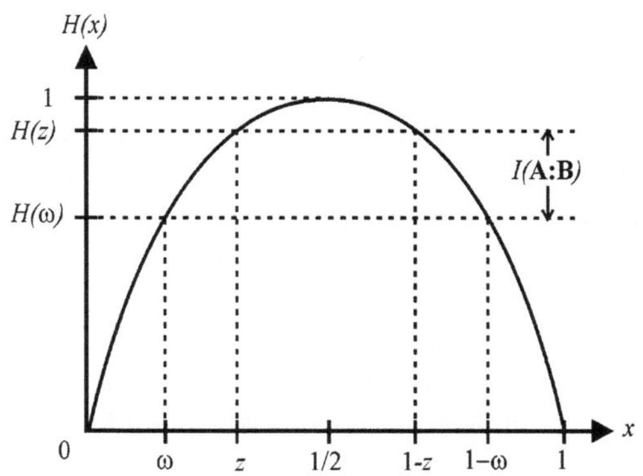

Fig. 4 The binary entropy function $H(x) = -x\log_2 x - (1 - x)\log_2(1 - x)$ and the geometric interpretation of the conditional entropy $S(\boldsymbol{q}|\boldsymbol{p}) = H(\omega) \equiv H(\mathbf{B}|\mathbf{A})$ and mutual information $I(\boldsymbol{p}:\boldsymbol{q}) = H(z) - H(\omega) \equiv I(\mathbf{A}:\mathbf{B})$ quantities for the information network of Fig. 3

system output entropy is also determined by another BEF, $S(\boldsymbol{q}) = H(z(x, \omega))$, where $z(x, \omega) \equiv q_2 = x\omega + (1 - x)(1 - \omega)$, while the channel conditional entropy $S(\boldsymbol{q}|\boldsymbol{p}) = H(\omega)$ measures its average communication noise. The mutual information between the system inputs and outputs, $I(\boldsymbol{p}:\boldsymbol{q}) = S(\boldsymbol{q}) - S(\boldsymbol{q}|\boldsymbol{p}) = H[z(x, \omega)] - H(\omega)$, thus reflects the system net information flow (see Fig. 4). Since z always lies between ω and $1 - \omega$, $H(z) = H(1 - z) \geq H(\omega) = H(1 - \omega)$. This demonstrates the *non*negative character of the mutual information, represented by the overlap areas between the two entropy circles in a qualitative diagram of Fig. 1.

The amount of information $I(\boldsymbol{p}:\boldsymbol{q})$ flowing through this transmission system thus depends on both the cross-over probability ω, which characterizes the communication channel itself, and on the input probability x, which determines the way the device is exploited (probed). For $x = 0$ (or 1) $H(z) = H(\omega)$ and thus $I(\boldsymbol{p}:\boldsymbol{q}) = 0$, i.e., there is no net flow of information from the source to the receiver. For $x = ½$ one finds $H(z) = 1$ bit, thus giving rise to the maximum value of the mutual information (transmission capacity): $C(\omega) \equiv \max_{\boldsymbol{p}} I(\boldsymbol{p}:\boldsymbol{q}) = \max_x\{H[z(x, \omega)] - H(\omega)\} = 1 - H(\omega)$. Hence, for $\omega = ½$ the information capacity of SBC identically vanishes.

4 Information Displacements in Molecules

The separated atoms (in their ground states), when shifted to their molecular positions define the (isoelectronic) molecular prototype called the "promolecule" [9, 25]. Its overall electron density, given by the sum of the free-atom distributions $\{\rho_i^0\}$, $\rho^0 = \sum_i \rho_i^0$, defines the initial stage in the bond-formation process. It determines a natural reference for extracting changes due to formation of chemical bonds. They are embodied in the *density-difference* function $\Delta\rho = \rho - \rho^0$, between the molecular ground-state density $\rho = \sum_i \rho_i$, a collection of the AIM densities $\{\rho_i\}$, and the promolecular electron distribution ρ^0. This deformation density has been widely used to probe the electronic structure of molecular systems. In this section we compare it against selected (local) IT probes, in order to explore the information content of the molecular ground-state electron distribution $\rho(\boldsymbol{r})$, or its shape (probability) factor $p(\boldsymbol{r}) = \rho(\boldsymbol{r})/N$ [9–16], obtained from the Kohn–Sham (KS) [65] calculations in the Local Density Approximation (LDA) of the Density Functional Theory (DFT) [66].

Consider the density $\Delta s(\boldsymbol{r})$ of the Kullback–Leibler functional for the information distance between the molecular and promolecular electron distributions,

$$\Delta S[\rho|\rho^0] = \int \rho(\boldsymbol{r})\ln[\rho(\boldsymbol{r})/\rho^0(\boldsymbol{r})]\mathrm{d}\boldsymbol{r} = \int \rho(\boldsymbol{r})I[w(\boldsymbol{r})]\mathrm{d}\boldsymbol{r} \equiv \int \Delta s(\boldsymbol{r})\mathrm{d}\boldsymbol{r}, \tag{13}$$

where $w(\boldsymbol{r})$ and $I[w(\boldsymbol{r})]$ denote the density-enhancement and surprisal functions relative to promolecule, respectively. We shall also explore the density $\Delta h_\rho(\boldsymbol{r})$ of the molecular displacement in the Shannon entropy,

$$\Delta H[\rho|\rho^0] = S[\rho] - S[\rho^0] \equiv -\int \rho(\boldsymbol{r})\ln\rho(\boldsymbol{r})\mathrm{d}\boldsymbol{r} + \int \rho^0(\boldsymbol{r})\ln\rho^0(\boldsymbol{r})\mathrm{d}\boldsymbol{r} \equiv \int \Delta h_\rho(\boldsymbol{r})\mathrm{d}\boldsymbol{r}, \tag{14}$$

testing its performance as an alternative IT tool for diagnosing the presence of chemical bonds and monitoring the effective valence states of bonded atoms.

It should be observed that the molecular electron density $\rho(\boldsymbol{r}) = \rho^0(\boldsymbol{r}) + \Delta\rho(\boldsymbol{r})$ is only slightly modified relative to the promolecular distribution $\rho^0(\boldsymbol{r})$: $\rho(\boldsymbol{r}) \approx \rho^0(\boldsymbol{r})$ or $w(\boldsymbol{r}) \approx 1$. Indeed, the formation of chemical bonds involves only a minor reconstruction of the electronic structure, mainly in the valence shells of the constituent atoms, so that $|\Delta\rho(\boldsymbol{r})| \equiv |\rho(\boldsymbol{r}) - \rho^0(\boldsymbol{r})| \ll \rho(\boldsymbol{r}) \approx \rho^0(\boldsymbol{r})$. Hence, the ratio $\Delta\rho(\boldsymbol{r})/\rho(\boldsymbol{r}) \approx \Delta\rho(\boldsymbol{r})/\rho^0(\boldsymbol{r})$ is generally small in the energetically important regions of large density values.

As explicitly shown in the first column of Fig. 5, the largest values of the density difference $\Delta\rho(\boldsymbol{r})$ are observed mainly in the bond region, between the nuclei of chemically bonded atoms. The accompanying reconstruction of atomic lone pairs is also seen to lead to an appreciable displacement in the molecular electron density. By expanding the logarithm of the molecular surprisal $I[w(\boldsymbol{r})]$ around $w(\boldsymbol{r}) = 1$, to the first-order in the relative displacement of the electron density, one obtains the following approximate relation between the local value of the molecular surprisal density and that of the difference function itself:

$$I[w(\boldsymbol{r})] = \ln[\rho(\boldsymbol{r})/\rho^0(\boldsymbol{r})] = \ln\{[\rho^0(\boldsymbol{r}) + \Delta\rho(\boldsymbol{r})]/\rho^0(\boldsymbol{r})\} \cong \Delta\rho(\boldsymbol{r})/\rho^0(\boldsymbol{r}) \\ \approx \Delta\rho(\boldsymbol{r})/\rho(\boldsymbol{r}). \tag{15}$$

It provides a semiquantitative IT interpretation of the relative density difference diagrams and links the local surprisal of IT to the density difference function of quantum chemistry. This equation also relates the integrand of the information-distance functional with the corresponding displacement in the electron density: $\Delta s(\boldsymbol{r}) = \rho(\boldsymbol{r})I[w(\boldsymbol{r})] \cong \Delta\rho(\boldsymbol{r})w(\boldsymbol{r}) \approx \Delta\rho(\boldsymbol{r})$.

This approximate relation is numerically verified in Fig. 5, where the contour diagram of the directed-divergence density $\Delta s(\boldsymbol{r})$ (second column) is compared with the corresponding map of its first-order approximation $\Delta\rho(\boldsymbol{r})w(\boldsymbol{r})$ (third column) and the density difference function itself (first column). A general similarity between the diagrams in each row of the figure confirms a semiquantitative character of this first-order expansion. A comparison between panels of the first two columns in the figure shows that the two displacement maps so strongly resemble one another that they are hardly distinguishable. This confirms a close relation between the local density and entropy-deficiency relaxation patterns, thus attributing to the former the complementary IT interpretation of the latter. The density displacement and the missing-information distribution can be thus viewed as practically equivalent probes of the system chemical bonds.

The density difference function $\Delta\rho(\boldsymbol{r})$ for representative linear diatomic and triatomic molecules reflects all typical reconstructions of free atoms in molecules, which accompany the formation of a single or multiple chemical bonds with varying degree of the *covalency* (electron *sharing*) and *ionicity* (*charge transfer*) components. For example, the single covalent bond in H_2 gives rise to a relative accumulation of electrons in the bond region, between the two nuclei, at the

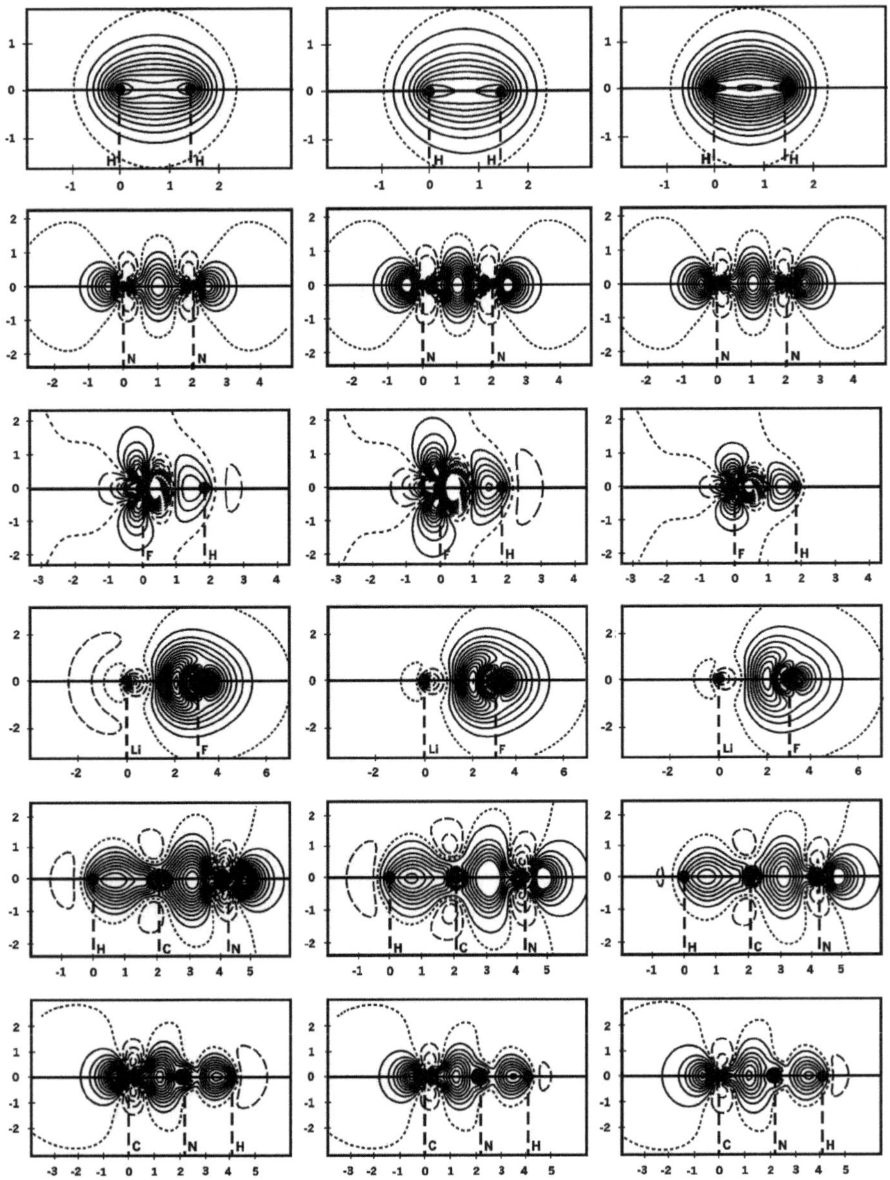

Fig. 5 The contour diagrams of the molecular density difference function $\Delta\rho(\boldsymbol{r}) = \rho(\boldsymbol{r}) - \rho^0(\boldsymbol{r})$ (first column), the information-distance density $\Delta s(\boldsymbol{r}) = \rho(\boldsymbol{r})I[w(\boldsymbol{r})]$ (second column), and its approximate, first-order expansion $\Delta s(\boldsymbol{r}) \cong \Delta\rho(\boldsymbol{r})w(\boldsymbol{r})$ (third column), for selected diatomic and linear triatomic molecules: H_2, HF, LiF, HCN, and HNC. *The solid, pointed and broken lines* denote the positive, zero, and negative values, respectively, of the equally spaced contours [15, 67]

expense of the outer, nonbonding regions of space. The triple-bond pattern for N_2 reflects density accumulation in the bonding region, due to both the σ and π bonds, and the accompanying increase in the density of the lone pairs on both nitrogen atoms, due to their expected *sp*-hybridization. One also observes a decrease in the electron density in the vicinity of the nuclei and an outflow of electrons from the $2p_\pi$ AO to their overlap area, a clear sign of their involvement in formation of the double π bond system. In triatomic molecules one identifies a strongly covalent pattern of the electron density displacements in the regions of the single N–H and C–H atoms. A typical buildup of the bond charge due to the multiple CN bonds in the two isomers HCN and HNC can be also observed. The increase in the lone-pair electron density on the terminal heavy atom, N in HCN and or C in HNC, can be also detected, thus confirming the expected *sp*-hybridization of these bonded atoms.

Both heteronuclear diatomics represent partially ionic bonds between the two atoms exhibiting a small and large difference in their electronegativity (chemical hardness) descriptors, respectively. A pattern of the density displacement in HF reflects a weakly ionic (strongly covalent) bond, while in LiF the two AIM are seen to be connected by the strongly ionic (weakly covalent) bond. Indeed, in HF one detects a common possession of the valence electrons by the two atoms, giving rise to the bond charge located between them, and hence a small H $\rightarrow$ F charge transfer. Accordingly, in LiF a substantial Li $\rightarrow$ F electron transfer is detected.

In Fig. 6 the contour maps of the entropy-displacement density $\Delta h_\rho(\boldsymbol{r})$ are compared with the corresponding density difference diagrams for representative linear molecules of Fig. 5. Again, $\Delta\rho$ and Δh_ρ diagrams for H_2 are seen to qualitatively resemble one another and the corresponding Δs map of Fig. 5. The main feature of Δh_ρ plot, an increase in the electron uncertainty between nuclei, is due to the inflow of electrons to this region. Again this covalent charge/entropy accumulation reflects the electron-sharing effect and the delocalization of the AIM valence electrons, which effectively move in the field of both nuclei, towards the bond partner. The entropy difference function is seen to display typical features in the reconstruction of atomic electron distributions in a molecule, relative to the promolecule.

Next we examine the central-bond problem in small [1.1.1], [2.1.1], [2.2.1], and [2.2.2] propellanes shown in Fig. 7. The main purpose of this study [9, 16] was to examine the effect on the central C′–C′ bond between the "bridgehead" carbon atoms of an increase in the size of the carbon bridges. The contour maps of Fig. 8 compare $\Delta\rho$, Δs, and Δh_ρ plots in the planes of section shown in Fig. 7 generated using the DFT–LDA calculations in the extended (DZVP) basis set. They display a depletion of the electron density between the bridgehead carbons in [1.1.1] and [2.1.1] propellanes, while larger bridges in [2.2.1] and [2.2.2] systems generate a density buildup in this region. A similar conclusion follows from the entropy-displacement and entropy-deficiency plots of the figure. The two entropic diagrams are again seen to be qualitatively similar to the corresponding density difference plots. As before, this resemblance is seen to be particularly strong between the $\Delta\rho$ and Δs diagrams.

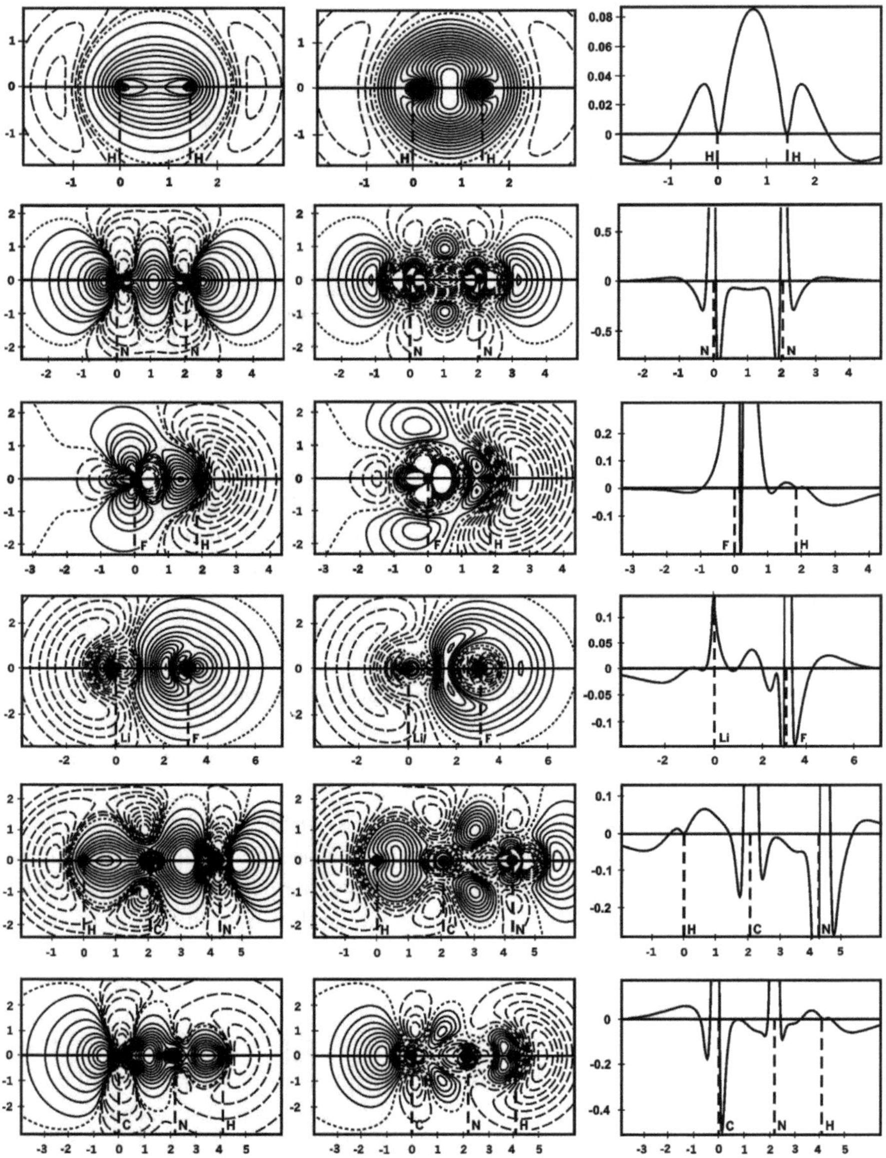

Fig. 6 Comparison between the nonequidistant contour diagrams of the density-difference $\Delta\rho(\boldsymbol{r})$ (first column) and entropy-difference $\Delta h_\rho(\boldsymbol{r})$ (second column) functions for the linear molecules of Fig. 5; third column shows the profiles of $\Delta h_\rho(\boldsymbol{r})$ for cuts along the bond axis [9, 16]

5 Contra-Gradience Probe of Bond Localization

Each density functional $A[\rho]$ can be regarded as the corresponding *multi*component functional $A[\rho] \equiv A^{\text{total}}[\boldsymbol{\rho}]$ of the pieces $\boldsymbol{\rho} = \{\rho_\alpha\}$ into which the electron distribution is decomposed:

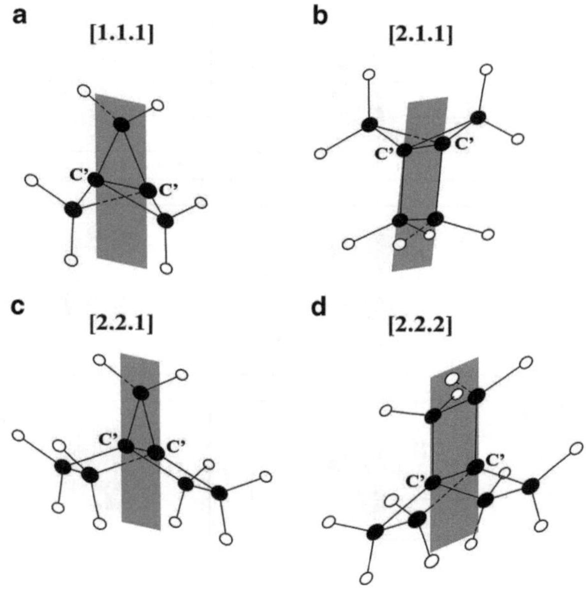

Fig. 7 The propellane structures and the planes of sections containing the bridge and bridgehead (C′) carbon atoms, all identified by *black circles*

$$\rho = \sum\nolimits_{\alpha} \rho_{\alpha}. \tag{16}$$

Such functionals appear in the non-Born–Oppenheimer DFT of molecules [68], in partitioning the electron density into distributions assigned to AIM, and in the fragment embedding problems [69–72]. Each resolution of the molecular electron density also implies the associated division of the molecular (*total*) quantity $A^{\mathrm{total}}[\boldsymbol{\rho}]$ into its *additive*, $A^{\mathrm{add.}}[\boldsymbol{\rho}]$, and *nonadditive*, $A^{\mathrm{nadd.}}[\boldsymbol{\rho}]$, contributions:

$$A^{\mathrm{add.}}[\boldsymbol{\rho}] = \sum\nolimits_{\alpha} A[\rho_{\alpha}],\ A^{\mathrm{nadd.}}[\boldsymbol{\rho}] = A^{\mathrm{total}}[\boldsymbol{\rho}] - A^{\mathrm{add.}}[\boldsymbol{\rho}]. \tag{17}$$

For example, the Gordon–Kim [73] type division of the kinetic energy functional defines the nonadditive contribution which constitutes the basis of the DFT-embedding concept of Cortona and Wesołowski [69–72]. Similar partition can be used to resolve the information quantities themselves. In particular, the inverse of the nonadditive Fisher information in the MO resolution has been shown to define the IT-ELF concept [29], in the spirit of the original Becke and Edgecombe formulation [30], while the related quantity in the AO resolution of the SCF MO theory generates the CG criterion for localizing the chemical bonds [9–14, 26–28]. It has been argued elsewhere that the valence basins of the negative CG density identify the bonding (constructive) interference of AO in the direct bonding mechanism, while the positive values of this local IT probe similarly delineate the antibonding regions in molecules (see Fig. 9).

For the given pair of AO, say $A(\boldsymbol{r})$ and $B(\boldsymbol{r})$ originating from different atoms A and B, respectively, the negative contribution to the nonadditive Fisher

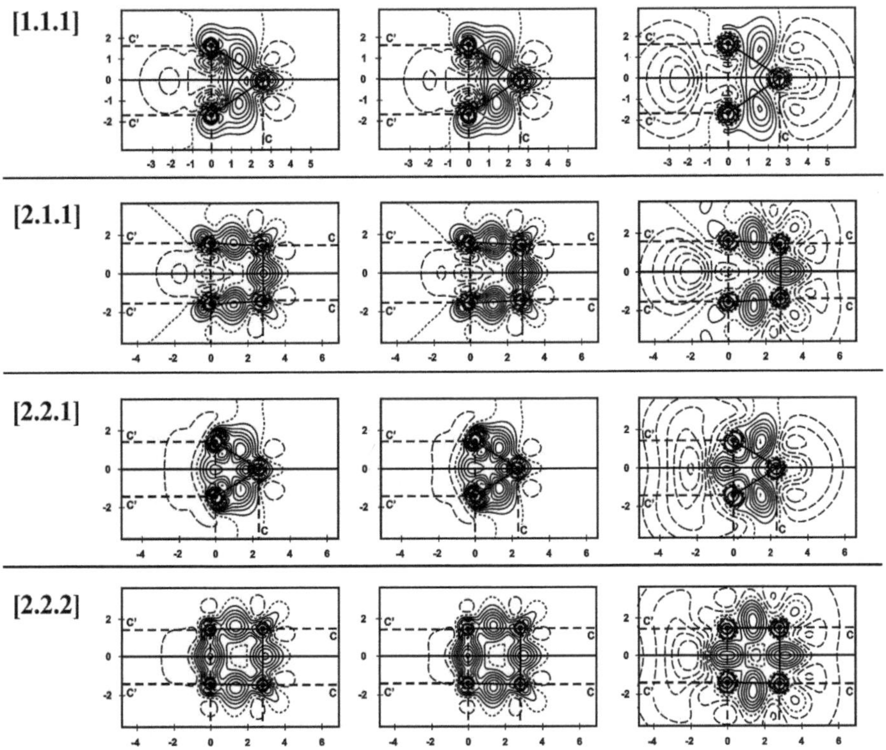

Fig. 8 A comparison between the (equidistant) contour maps of the density-difference function $\Delta\rho(\boldsymbol{r})$ (first column), the information-distance density $\Delta s(\boldsymbol{r})$ (second column), and the entropy-displacement density $\Delta h_\rho(\boldsymbol{r})$ (third column), for the four propellanes of Fig. 7 [9, 16]

information results, when the gradient of one AO exhibits the negative projection on the direction of the gradient of the other AO. This explains the name of the CG criterion itself. The zero contour, which encloses the bonding region and separates it from the antibonding environment, is thus identified by the equation $i^{cg}(\boldsymbol{r}) \equiv \nabla A(\boldsymbol{r}) \cdot \nabla B(\boldsymbol{r}) = 0$. As shown in the qualitative diagram of Fig. 9, for the two 1 s orbitals in H_2 this dividing surface constitutes the sphere passing through both nuclei, on which the two AO gradients are mutually perpendicular. Integration of i^{cg} over the whole space gives the associated CG integral,

$$I^{cg} = \int i^{cg}(\boldsymbol{r})\mathrm{d}\boldsymbol{r} = \int \nabla A(\boldsymbol{r}) \cdot \nabla B(\boldsymbol{r})\mathrm{d}\boldsymbol{r} = -\int A(\boldsymbol{r})\,\Delta B(\boldsymbol{r})\mathrm{d}\boldsymbol{r} = \frac{2\mu}{\hbar^2}\langle A|\,\hat{\mathrm{T}}|B\rangle \equiv \frac{2\mu}{\hbar^2}T_{A,B}, \tag{18}$$

where μ stands for the electronic mass, while limiting this integration to the valence (bonding) region of the negative i^{cg}, inside the closed surface $i^{cg}(\boldsymbol{r}) = 0$ (see Fig. 9), provides a useful tool for indexing the chemical bonds [28]. As also indicated in the preceding equation, the CG integral measures the coupling (off-diagonal) element

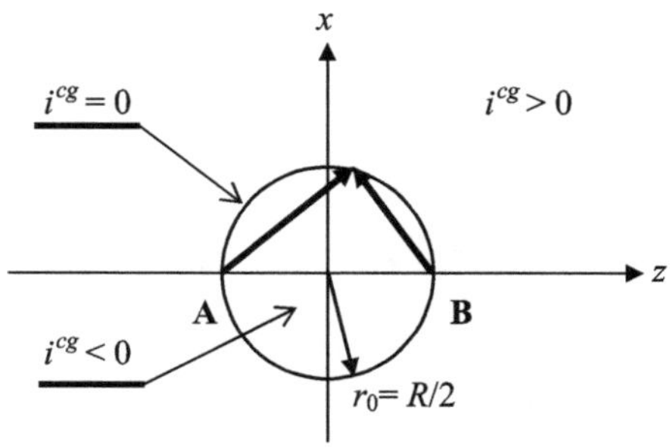

Fig. 9 The circular contour passing through both nuclei of the vanishing CG integrand for two 1*s* orbitals on atoms A and B, $i^{cg}(\boldsymbol{r}) = 0$. It separates the bonding region (inside the circle), where $i^{cg}(\boldsymbol{r}) < 0$, from the region of positive contributions $i^{cg}(\boldsymbol{r}) > 0$ (outside the circle). One observes that the *heavy arrows* representing the negative gradients of the two basis functions are mutually perpendicular on the dividing (spherical) surface $i^{cg}(\boldsymbol{r}) = 0$

$T_{A,B}$ of the electronic kinetic energy operator $\hat{\mathrm{T}}$. Thus, CG integral reflects the kinetic energy coupling between the two chemically interacting AO. Such integrals are routinely calculated in typical quantum-chemical packages for determining the electronic structure of molecular systems. This observation emphasizes the crucial role of the kinetic energy component in interpreting the IT origins of the chemical bonding [74–77].

Consider a general case of the AO basis set $\boldsymbol{\chi} = (\chi_1, \chi_2, \ldots, \chi_m)$ used to describe the occupied MO in the *N*-electron system. In the ground-state electron configuration defined by the *singly* occupied subspace $\boldsymbol{\psi} = \{\psi_k\}$ of the *N* lowest *spin*-MO, with the spatial (MO) parts $\boldsymbol{\varphi} = \boldsymbol{\chi}\mathbf{C} = \{\varphi_k,\ k = 1, 2, \ldots, N\}$ generated by the Hartree–Fock (SCF MO) or KS calculations, the nonadditive Fisher information in AO resolution reads as follows:

$$I^{\text{nadd.}}[\boldsymbol{\chi}] = 4\sum_{k=1}^{m}\sum_{l=1}^{m}\int \gamma_{k,l}(1-\delta_{k,l})\nabla\chi_l^*(\boldsymbol{r})\cdot\nabla\chi_k(\boldsymbol{r})\mathrm{d}\boldsymbol{r} \equiv 2\int f^{\text{nadd.}}(\boldsymbol{r})\mathrm{d}\boldsymbol{r} = 8T^{\text{nadd.}}[\boldsymbol{\chi}]. \tag{19}$$

Here, the elements of the *Charge-and-Bond-Order* (CBO), first-order density matrix,

$$\boldsymbol{\gamma} = \left\langle \boldsymbol{\chi} \left| \left(\sum_{k=1}^{N} |\varphi_k\rangle\langle\varphi_k| \right) \right| \boldsymbol{\chi} \right\rangle = \langle\boldsymbol{\chi}|\boldsymbol{\varphi}\rangle\langle\boldsymbol{\varphi}|\boldsymbol{\chi}\rangle \equiv \langle\boldsymbol{\chi}|\hat{\mathbf{P}}_{\varphi}|\boldsymbol{\chi}\rangle = \mathbf{C}\mathbf{C}^{\dagger} = \{\gamma_{u,w}\}, \tag{20}$$

provide the AO representation of the projection operator $\hat{\mathrm{P}}_{\varphi}$ onto the occupied MO subspace. The average nonadditive information is thus proportional to the associated component $T^{\text{nadd.}}[\boldsymbol{\chi}]$ of the system average kinetic energy: $T^{\text{total}}[\boldsymbol{\chi}] = \mathrm{tr}(\boldsymbol{\gamma}\mathbf{T}) = T^{\text{add.}}[\boldsymbol{\chi}] + T^{\text{nadd.}}[\boldsymbol{\chi}]$, where the kinetic energy matrix in AO representation $\mathbf{T} = \{T_{k,l} = \langle\chi_k|\hat{\mathrm{T}}|\chi_l\rangle\}$ and $T^{\text{add.}}[\boldsymbol{\chi}] = \sum_k \gamma_{k,k} T_{k,k}$.

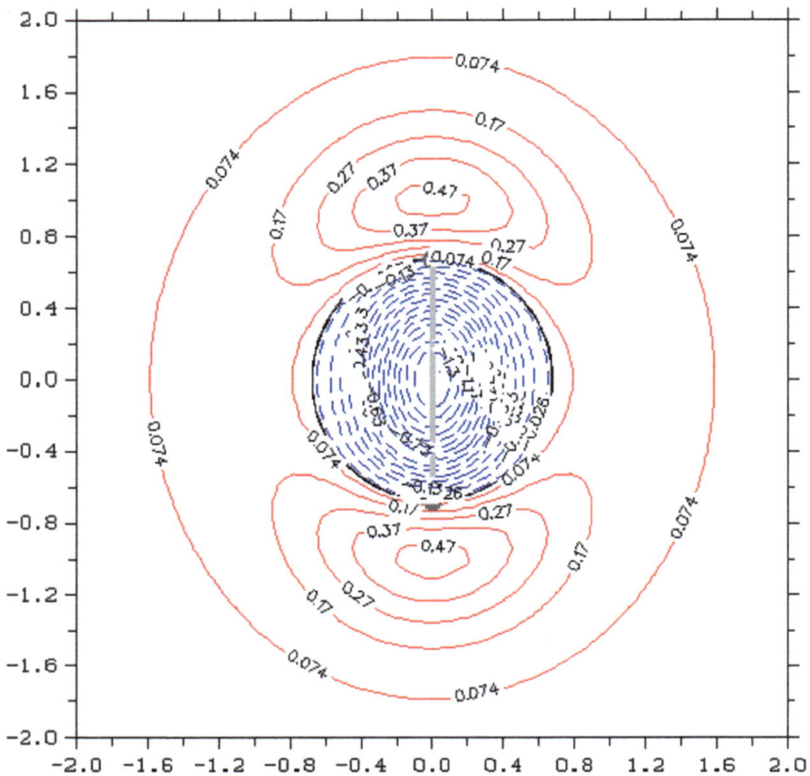

Fig. 10 The contour map of the CG density $f^{\text{nadd.}}(\boldsymbol{r})$ for H_2

In this orbital approximation one uses the most extended (valence) basins of the negative CG density, $f^{\text{nadd.}}(\boldsymbol{r}) < 0$, enclosed by the associated $f^{\text{nadd.}}(\boldsymbol{r}) = 0$ surfaces, as locations of the system chemical bonds. This probe has been successfully validated in recent numerical calculations [28]. In the remaining part of this section, we shall present selected results of this study. They have been obtained using the standard SCF MO calculations (GAMESS software) in the minimum (STO-3 G) Gaussian basis set. The contour maps, for the optimized geometries, are reported in atomic units. In these plots the broken-line contours correspond to the negative CG values. For visualization purposes representative perspective views of the CG bonding regions are also presented.

The contour map of Fig. 10 confirms qualitative predictions of Fig. 9 for H_2. In the minimum basis set one indeed observes a spherical CG bonding region between the two nuclei, with the accompanying increases of this quantity being observed in the nonbonding surroundings. A similar analysis for HF is presented in Fig. 11. It indicates existence of three basins of the negative CG density: a large valence region between the two atoms, and two small volumes in the inner shell of F. In the triple-bond case of N_2 (Fig. 12) the bonding basin is now distinctly extended away from the bond axis, due to the presence of the two π bonds

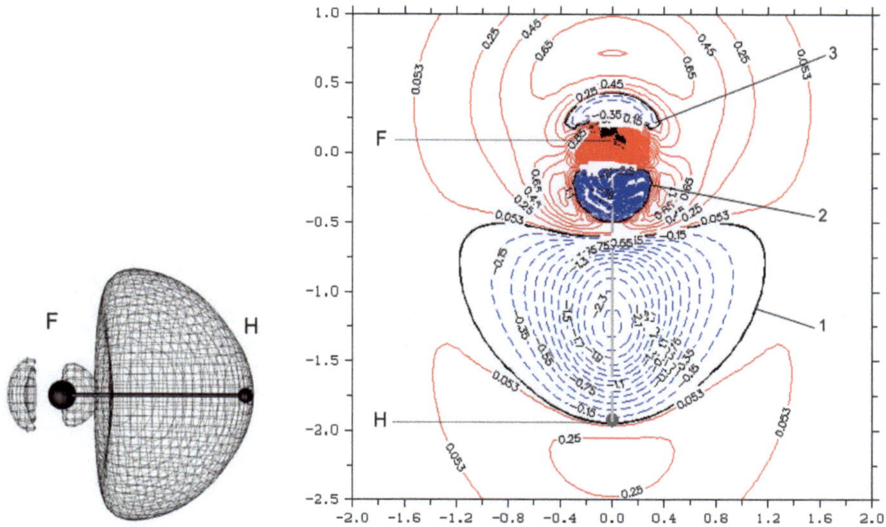

Fig. 11 The perspective view of the negative CG basins and the contour map of $f^{\text{nadd.}}(\boldsymbol{r})$ for HF

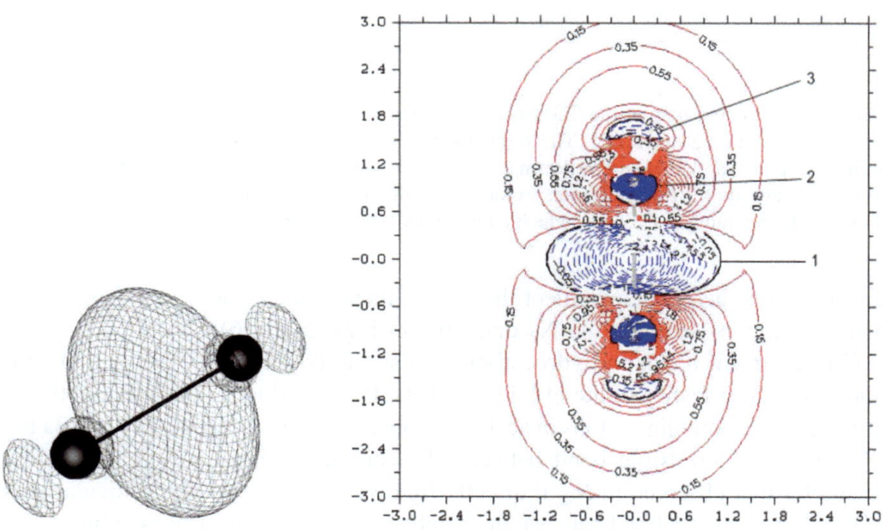

Fig. 12 The same as in Fig. 11 for N_2

accompanying the central σ bond. Small core-polarization basins near each nucleus are again seen in the perspective plot. The *sp*-hybridization promotion of the nonbonding regions on both atoms is again much in evidence in the accompanying contour map, and the bonding region is seen to be "squeezed" between the two atomic cores. The positive values of CG density in transverse

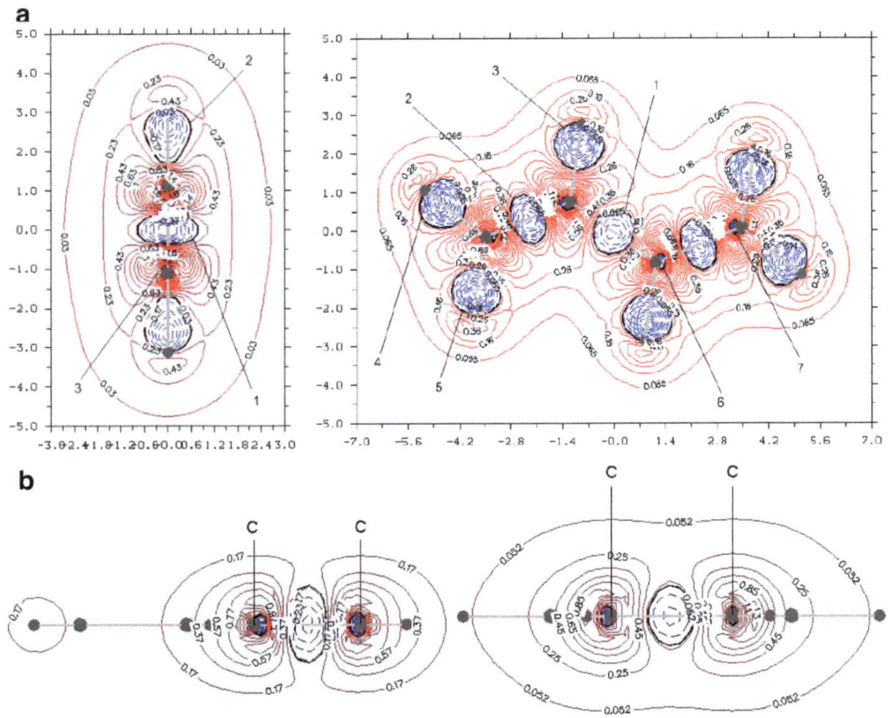

Fig. 13 The contour map of the CG density $f^{\mathrm{nadd.}}(\boldsymbol{r})$ for acetylene (left diagram of panel *a*) and butadiene (remaining diagrams). The right diagram of panel *a* shows the contour map in the molecular plane of butadiene, reflecting only the σ-bonds, while panel *b* reports additional, perpendicular cuts for this molecule, passing through terminal (lower-left diagram) and central (lower-right diagram) C–C bonds, which additionally reflect the π-bonds

directions on each nucleus reflect the charge displacements accompanying the π-bond formation (see also the N_2 diagrams in Figs. 5 and 6).

The chemical bonds in hydrocarbons are similarly probed in Figs. 13 and 14. These diagrams testify to the efficiency of the CG criterion in localizing both the C–C and C–H bonding regions in acetylene, butadiene, and benzene. In Fig. 13a the CG pattern of the triple bond between the carbon atoms in acetylene strongly resembles that observed in N_2. It directly follows from the two perpendicular cuts of Fig. 13b that the π bond between the neighboring peripheral carbons in butadiene is stronger than its central counterpart. One thus concludes that the CG probe of chemical bonds indeed provides an efficient tool for locating the directly bonded regions in typical diatomic and polyatomic molecules.

Finally, the bonding patterns in a series of four small propellanes of Fig. 7 are examined in the contour maps of Fig. 15. Each row of this figure is devoted to a different propellane in a series: [1.1.1], [2.1.1], [2.2.1], [2.2.2]. The left panels of each row correspond to the plane of section perpendicular to the central bond between the bridgehead carbons, at the bond midpoint, while the axial cuts of the

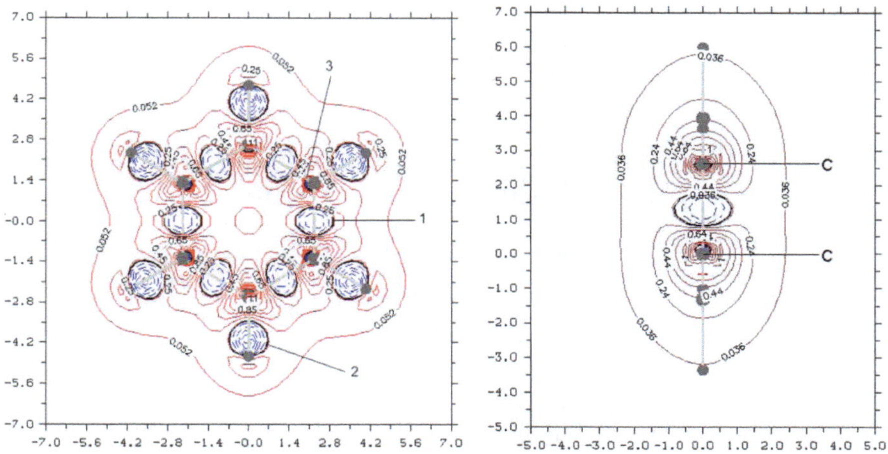

Fig. 14 The contour maps of the CG density $f^{\text{nadd.}}(\boldsymbol{r})$ for benzene: in the molecular plane (*left panel*) and in the perpendicular plane passing through the C–C bond (*right panel*)

right panels involve one of the carbon bridges. The main result of the density-difference and entropy-deficiency analysis of Fig. 8, of an apparent absence of the direct (through-space) bond between the carbon bridgeheads in the [1.1.1] and [2.1.1] systems and its full presence in the [2.2.1] and [2.2.2] propellanes, remains generally confirmed by the CG probe of Fig. 15. However, this transition is now seen to be less sharp, with very small bonding basins between bridgeheads being also observed in the two smallest molecules. This small bonding basin of [1.1.1] system is seen to gradually evolve into that attributed to the full bond in the [2.2.2] propellane. The bridge C–C and C–H chemical bonds are again seen to be perfectly localized by the closed valence surfaces of the vanishing CG density.

One of the primary goals of theoretical chemistry is to identify also the physical sources of the chemical bond. Most of existing theoretical interpretations of its origins emphasize, almost exclusively, the potential (interaction) aspect of this phenomenon, focusing on the mutual attraction between the accumulation of electrons between the two atoms (the negative "bond charge") and the partially screened (positively charged) nuclei. This is indeed confirmed by the virial-theorem decomposition of the diatomic Born–Oppenheimer potentials. The latter indicates that for the equilibrium bond length it is the change in its potential component, relative to the dissociation limit, which is ultimately responsible for the net bonding effect of the atomic interaction.

The ELF and CG criteria adopt the complementary view, stressing the importance of the kinetic-energy bond component in locating the bonding regions in molecules. We recall that the associated displacement of the total kinetic energy of electrons, relative to that in the separated atom limit, is bonding only at an earlier stage of the mutual approach by both atoms. At this stage it is dominated by the *longitudinal* contribution associated with the gradient component along the bond axis. It then assumes the destabilizing character at the equilibrium internuclear

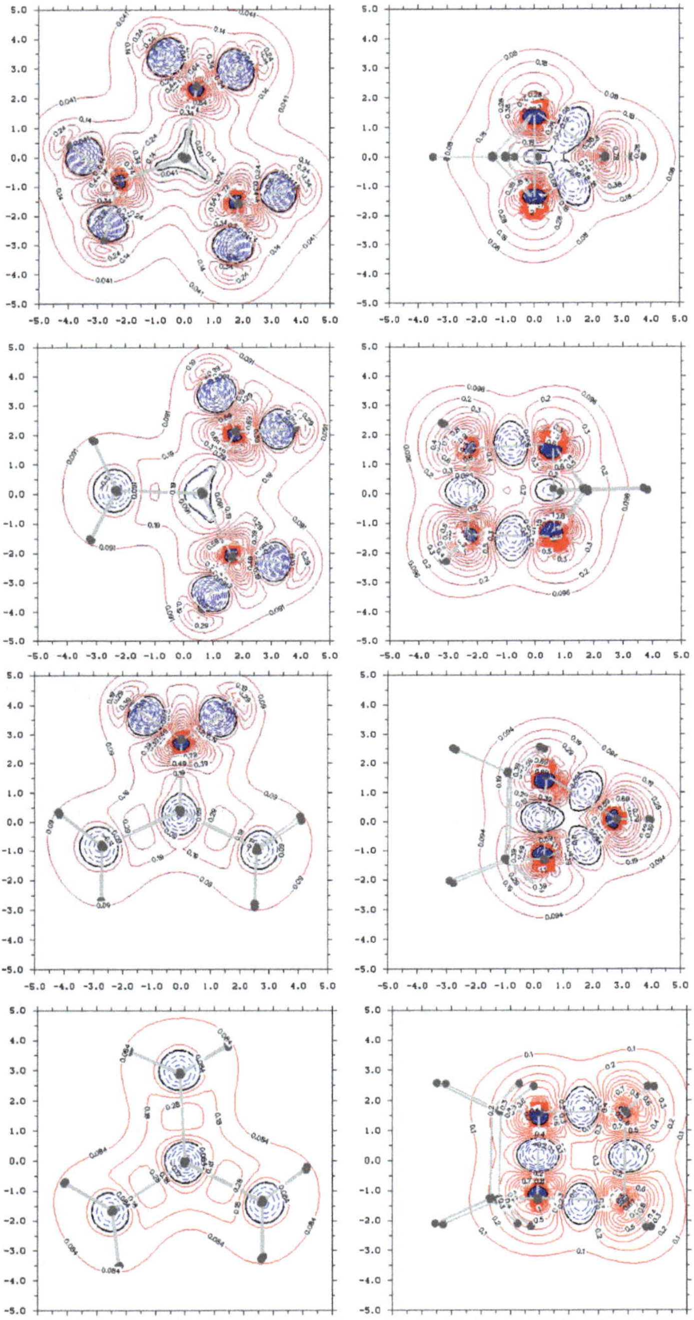

Fig. 15 The same as in Fig. 10 for the four propellanes of Fig. 7

separation, mainly due to its *transverse* contribution due to the gradient components in the directions perpendicular to the bond axis [74, 76, 77]. This virial theorem perspective also indicates that the kinetic energy constitutes a driving force of the bond-formation process. It follows from the classical analysis by Ruedenberg and coworkers [74, 76, 77] that a contraction of atomic electron distributions is possible in molecule due to this very lowering of the kinetic energy at large internuclear separation. The process of redistributing electrons in the chemical bond formation can be thus regarded as being "catalyzed" by the gradient effect of the kinetic energy. A similar conclusion follows from the theoretical analysis by Goddard and Wilson [75].

Therefore, the overall change in the kinetic energy (Fisher information) due to the chemical bond formation emphasizes a contraction of the electronic density in the presence of the remaining nuclear attractor in the molecule. It only blurs the subtle (*interference*) origins of chemical bonds. Indeed, the *total* kinetic energy component combines delicate, truly bonding (nonadditive) *interatomic* effects, originating from the stabilizing combinations of AO in the occupied MO, and the accompanying processes of the *intraatomic* polarization. In other words, the overall displacement in the kinetic energy contribution effectively hides a contribution due to minute changes in the system valence shell, which are associated by chemists with the chemical bond concept. Some partitioning of this overall energy component is thus called for to separate these subtle bonding phenomena from the associated nonbonding promotion of constituent atoms. Only by focusing on the nonadditive part of the electronic kinetic energy in CG criterion can one uncover the real information origins of the chemical bond and ultimately define efficient IT probe for its localization.

6 Orbital Communications and Information Bond Multiplicities

The molecular information system [9–13, 33, 39, 48, 55–58] represents the key concept of CTCB. It can be constructed at alternative levels defined by the underlying electron-localization "events," which determine the channel inputs $\boldsymbol{a} = \{a_i\}$ and outputs $\boldsymbol{b} = \{b_j\}$. In OCT the AO basis functions of the SCF MO calculations determine a natural resolution for discussing the information multiplicity (order) of the system chemical bonds: $\boldsymbol{a} = \{\chi_i\}$ and $\boldsymbol{b} = \{\chi_j\}$. This AO network describes the probability/information propagation in the molecule. It can be described by the standard quantities developed in IT for real communication devices. The transmission of the AO-assignment "signals" becomes randomly disturbed, due to the electron delocalization throughout the network of chemical bonds, thus exhibiting typical communication "noise." Indeed, an electron initially attributed to the given AO in the channel "input" can be later found with a nonvanishing probability at several AO in the molecular "output."

This electron delocalization is embodied in the conditional probabilities of the "outputs-given-inputs," $\mathbf{P}(\boldsymbol{b}|\boldsymbol{a}) = \{P(\chi_j|\chi_i) \equiv P(j|i)\}$, which define the "forward" channel of orbital communications. In OCT one constructs these probabilities [39, 48, 55–58, 78] using the bond-projected superposition principle of quantum mechanics [79]. This "physical" projection involves all occupied MO, which ultimately determine the entire network of chemical bonds in the molecular system of interest. Both the molecule as a whole and its constituent subsystems can be adequately described in terms of such IT bond multiplicities [9–13]. The *off*-diagonal orbital communications are related to the familiar Wiberg [80] contributions to the molecular bond orders or the related "quadratic" bond multiplicities [81–90] formulated in the MO theory.

The IT descriptors of the chemical bond pattern have been shown to account for the chemical intuition quite well providing the resolution of the diatomic bond multiplicities into the complementary IT-covalent and IT-ionic components [10, 11, 56]. The *internal* (intrafragment) and *external* (interfragment) indices of molecular subsystems (groups of AO) can be efficiently generated using the appropriate *reduction* of the molecular AO channel by combining the selected outputs into larger molecular fragments [9, 37, 42, 78].

In the SCF MO theory the bond network is determined by the occupied MO in the system ground state. For reasons of simplicity we assume the closed-shell (*cs*) configuration of $N = 2n$ electrons in the standard (spin-restricted) Hartree–Fock (RHF) description, which involves n lowest (doubly occupied, orthonormal) MO. In the familiar LCAO MO approach they are expanded as linear combinations of the (Löwdin) orthogonalized AO $\boldsymbol{\chi} = (\chi_1, \chi_2, \ldots, \chi_m) = \{\chi_i\}$ contributed by the system constituent atoms:

$$\langle\boldsymbol{\chi}|\boldsymbol{\chi}\rangle = \{\delta_{i,j}\} \equiv \mathbf{I}, \quad \boldsymbol{\varphi} = (\varphi_1, \varphi_2, \ldots, \varphi_n) = \{\varphi_s\} = \boldsymbol{\chi}\mathbf{C};$$

here, the rectangular matrix $\mathbf{C} = \{C_{i,s}\} = \langle\boldsymbol{\chi}|\boldsymbol{\varphi}\rangle$ groups the relevant LCAO MO coefficients to be determined using the iterative self-consistent-field procedure.

The molecular electron density,

$$\rho(\boldsymbol{r}) = 2\boldsymbol{\varphi}(\boldsymbol{r})\boldsymbol{\varphi}^{\dagger}(\boldsymbol{r}) = \boldsymbol{\chi}(\boldsymbol{r})[2\mathbf{C}\mathbf{C}^{\dagger}]\boldsymbol{\chi}^{\dagger}(\boldsymbol{r}) \equiv \boldsymbol{\chi}(\boldsymbol{r})\boldsymbol{\gamma}\boldsymbol{\chi}^{\dagger}(\boldsymbol{r}) = Np(\boldsymbol{r}), \tag{21}$$

and hence also the probability distribution $p(\boldsymbol{r}) = \rho(\boldsymbol{r})/N$, the *shape*-factor of ρ, are both determined by the CBO matrix,

$$\boldsymbol{\gamma} = 2\langle\boldsymbol{\chi}|\boldsymbol{\varphi}\rangle\langle\boldsymbol{\varphi}|\boldsymbol{\chi}\rangle = 2\mathbf{C}\mathbf{C}^{\dagger} \equiv 2\langle\boldsymbol{\chi}|\hat{\mathrm{P}}_{\varphi}|\boldsymbol{\chi}\rangle = \{\gamma_{i,j} = 2\langle\chi_i|\hat{\mathrm{P}}_{\varphi}|\chi_j\rangle \equiv 2\langle i|\hat{\mathrm{P}}_{\varphi}|j\rangle\}. \tag{22}$$

The latter constitutes the AO representation of the projection operator onto the subspace of all (doubly) occupied MO, $\hat{\mathrm{P}}_{\varphi} = |\boldsymbol{\varphi}\rangle\langle\boldsymbol{\varphi}| = \sum_s|\varphi_s\rangle\langle\varphi_s| \equiv \sum_s \hat{P}_s$, and satisfies the idempotency relation:

$$(\boldsymbol{\gamma})^2 = 4\langle\boldsymbol{\chi}|\hat{\mathrm{P}}_{\varphi}|\boldsymbol{\chi}\rangle\langle\boldsymbol{\chi}|\hat{\mathrm{P}}_{\varphi}|\boldsymbol{\chi}\rangle = 4\langle\boldsymbol{\chi}|\hat{\mathrm{P}}_{\varphi}^2|\boldsymbol{\chi}\rangle = 4\langle\boldsymbol{\chi}|\hat{\mathrm{P}}_{\varphi}|\boldsymbol{\chi}\rangle = 2\boldsymbol{\gamma}. \tag{23}$$

The CBO matrix reflects the promoted, *valence* state of AO in the molecule. Its diagonal elements reflect the effective electron occupations of the basis functions, $\{N_i = \gamma_{i,i} = Np_i\}$, with the normalized probabilities $\boldsymbol{p} = \{p_i = \gamma_{i,i}/N\}$ of the basis functions occupancy in molecule: $\sum_i p_i = 1$.

The orbital information system involves the AO events in the channel input $\boldsymbol{a} = \{\chi_i\}$ and output $\boldsymbol{b} = \{\chi_j\}$. In this description the AO $\to$ AO communication network is determined by the conditional probabilities:

$$\mathbf{P}(\boldsymbol{b}|\boldsymbol{a}) = \{P(j|i) = P(i \wedge j)/p_i\}, \quad \sum_j P(j|i) = 1, \tag{24}$$

where the joint probabilities of simultaneously observing two AO in the system chemical bonds $\mathbf{P}(\boldsymbol{a}\wedge\boldsymbol{b}) = \{P(i\wedge j)\}$ exhibit the usual normalization relations:

$$\sum_i P(i \wedge j) = p_j, \quad \sum_j P(i \wedge j) = p_i, \quad \sum_i \sum_j P(i \wedge j) = 1. \tag{25}$$

These probabilities involve squares of corresponding elements of the CBO matrix [39, 48]:

$$\mathbf{P}(\boldsymbol{b}|\boldsymbol{a}) = \{P(j|i) = R_i \left|\langle i|\hat{\mathrm{P}}_{\varphi}|j\rangle\right|^2 = (2\gamma_{i,i})^{-1}\gamma_{i,j}\gamma_{j,i}\}, \tag{26}$$

where the constant $R_i = (2\gamma_{i,i})^{-1}$ satisfies the normalization condition of Eq. (24). These probabilities explore the dependencies between AO resulting from their participation in the framework of the occupied MO, i.e., their simultaneous involvement in the entire network of chemical bonds in the molecule. This orbital channel can be subsequently probed using both the promolecular ($\boldsymbol{p}^0 = \{p_i^{\,0}\}$), molecular ($\boldsymbol{p} = \{p_i\}$), or general (ensemble) input probabilities, in order to extract the desired aspects of the chemical bond pattern [10, 11, 56, 63].

The off-diagonal conditional probability of *j*th AO output given *i*th AO input is thus proportional to the squared element of the CBO matrix linking the two AO, $\gamma_{j,i} = \gamma_{i,j}$, thus being also proportional to the corresponding AO contribution $M_{i,j} = \gamma_{i,j}^{\;2}$ to Wiberg's [80] index of the overall chemical bond order between two atoms in the molecule,

$$M_{\mathrm{A,B}} = \sum_{i \in A} \sum_{j \in \mathrm{B}} M_{i,j}, \tag{27}$$

or to related quadratic descriptors of molecular bond multiplicities [81–90].

In OCT the entropy/information indices of the covalent/ionic components of chemical bonds represent complementary descriptors of the average communication noise and the amount of information flow in the molecular information channel [9–13, 33, 63]. One observes that the molecular input $\boldsymbol{P}(\boldsymbol{a}) \equiv \boldsymbol{p}$ generates the same distribution in the output of the molecular channel,

$$\boldsymbol{q} = \boldsymbol{p}\,\mathbf{P}(\boldsymbol{b}|\boldsymbol{a}) = \left\{\sum_i p_i P(j|i) \equiv \sum_i P(i \wedge j) = p_j\right\} = \boldsymbol{p}, \tag{28}$$

thus identifying $\boldsymbol{p}$ as the *stationary* probability vector of the molecular ground state. This purely molecular channel is devoid of any reference (history) of the chemical bond formation and generates the average-noise index of the molecular overall IT covalency. It is measured by the conditional entropy of the system outputs given inputs:

$$S(\boldsymbol{\chi}|\boldsymbol{\chi}) \equiv S(\boldsymbol{P}(\boldsymbol{a})|\boldsymbol{P}(\boldsymbol{b})) = S(\boldsymbol{p}|\boldsymbol{p}) = -\sum_i \sum_j P(i \wedge j) \log[P(i \wedge j)/p_i] \equiv S. \quad (29)$$

The AO channel with the promolecular input "signal," $\boldsymbol{P}(\boldsymbol{a}^0) = \boldsymbol{p}^0$, refers to the initial state in the bond formation process. It corresponds to the ground state (fractional) occupations of the AO contributed by the system constituent (free) atoms, before their mixing into MO. These input probabilities give rise to the average information flow descriptor of the bond IT ionicity, given by the mutual information in the channel inputs and outputs [63]:

$$\begin{aligned} I^0(\boldsymbol{\chi} : \boldsymbol{\chi}) &\equiv I(\boldsymbol{P}(\boldsymbol{a}^0) : \boldsymbol{P}(\boldsymbol{b})) = I(\boldsymbol{p}^0 : \boldsymbol{p}) \\ &= \sum_i \sum_j P(i \wedge j) \log[P(i \wedge j)/(p_j p_i{}^0)] = S(\boldsymbol{p}) + \Delta S(\boldsymbol{p}|\boldsymbol{p}^0) - S \equiv I^0. \quad (30) \end{aligned}$$

In particlular, for the molecular input, when $\boldsymbol{p}^0 = \boldsymbol{p}$ and hence the vanishing information distance $\Delta S(\boldsymbol{p}|\boldsymbol{p}^0) = 0$, $I(\boldsymbol{p}{:}\boldsymbol{p}) = S(\boldsymbol{p}) - S \equiv I$.

The sum of these two bond components,

$$V^0(\boldsymbol{\chi}; \boldsymbol{\chi}) \equiv V(\boldsymbol{P}(\boldsymbol{a}^0); \boldsymbol{P}(\boldsymbol{b})) = V(\boldsymbol{p}^0; \boldsymbol{p}) = S + I^0 = S(\boldsymbol{p}) + \Delta S(\boldsymbol{p}|\boldsymbol{p}^0) \equiv V^0, \quad (31)$$

measures the overall IT bond multiplicity of all bonds in the molecular system under consideration. For the molecular input this quantity preserves the Shannon entropy of the molecular input probabilities:

$$V(\boldsymbol{p}; \boldsymbol{p}) = S(\boldsymbol{p}|\boldsymbol{p}) + I(\boldsymbol{p} : \boldsymbol{p}) = S(\boldsymbol{p}) \equiv V. \quad (32)$$

As an illustration consider the familiar problem of combining the two (Löwdin-orthogonalized) AO, $A(\boldsymbol{r})$ and $B(\boldsymbol{r})$, say, two 1s orbitals centered on nuclei A and B, respectively, which contribute a single electron each to form the chemical bond A–B. The two basis functions $\boldsymbol{\chi} = (A, B)$ then form the bonding (φ_b) and antibonding (φ_a) MO combinations, $\boldsymbol{\varphi} = (\varphi_b, \varphi_a) = \boldsymbol{\chi}\mathbf{C}$:

$$\varphi_b = \sqrt{P}A + \sqrt{Q}B, \quad \varphi_a = -\sqrt{Q}A + \sqrt{P}B, \quad P + Q = 1, \quad (33)$$

where the square matrix

$$\mathbf{C} = \begin{bmatrix} \sqrt{P} & -\sqrt{Q} \\ \sqrt{Q} & \sqrt{P} \end{bmatrix} = [\boldsymbol{C}_b | \boldsymbol{C}_a] \quad (34)$$

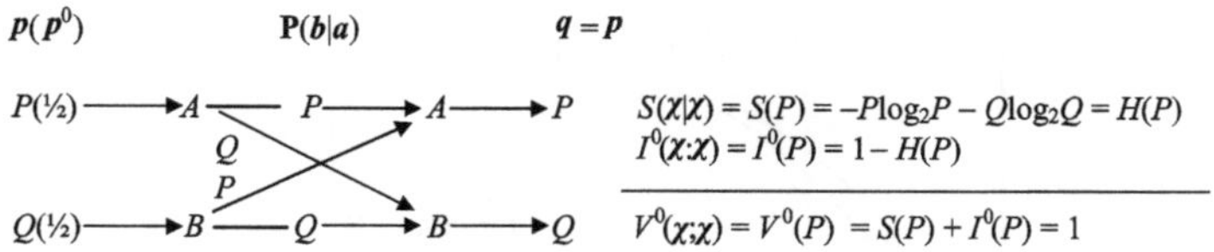

Fig. 16 Communication channel of the 2-AO model of the chemical bond and its entropy/information descriptors in bits

groups the LCAO MO expansion coefficients expressed in terms of the complementary probabilities: P and $Q = 1 - P$. They represent the conditional probabilities of observing AO in MO:

$$P(A|\varphi_b) = |C_{A,b}|^2 = P(B|\varphi_a) = |C_{B,a}|^2 \equiv P,$$
$$P(B|\varphi_b) = |C_{B,b}|^2 = P(A|\varphi_a) = |C_{A,a}|^2 = Q.$$

In the ground-state configuration of the doubly occupied bonding MO φ_b the CBO matrix $\boldsymbol{\gamma} = 2\mathbf{d}$, the double density matrix $\mathbf{d} = \langle \boldsymbol{\chi}|\varphi_b\rangle\langle\varphi_b|\boldsymbol{\chi}\rangle$, reads:

$$\boldsymbol{\gamma} = \{\gamma_{i,j}\} = 2\mathbf{C}_b\mathbf{C}_b{}^{\dagger} = 2\begin{bmatrix} P & \sqrt{PQ} \\ \sqrt{QP} & Q \end{bmatrix} \equiv 2\mathbf{d}. \tag{35}$$

It generates the following conditional probabilities of communications between AO in the molecular bond system,

$$\mathbf{P}(\boldsymbol{b}|\boldsymbol{a}) = \mathbf{P}(\boldsymbol{\chi}|\boldsymbol{\chi}) = \{P(j|i) = \gamma_{i,j}\gamma_{i,j}/(2\gamma_{i,i})\} = \begin{bmatrix} P & Q \\ P & Q \end{bmatrix}, \tag{36}$$

which determine the $\boldsymbol{\chi} \rightarrow \boldsymbol{\chi}$ communication network shown in Fig. 16.

This nonsymmetrical binary channel adopts the molecular input signal $\boldsymbol{p} = (P, Q)$ to extract the bond IT covalency, measuring the channel average communication noise, and the promolecular input signal $\boldsymbol{p}^0 = (½, ½)$, in which the two basis functions contribute a single electron each to form the chemical bond, in order to determine the model IT-ionicity index measuring the channel information capacity. The bond IT covalency $S(P)$ is determined by the binary entropy function $H(P)$ reaching the maximum value $H(P = ½) = 1$ bit for the symmetric bond $P = Q = ½$, e.g., the σ bond in H_2 or π-bond in ethylene (see also Figs. 4 and 17). It vanishes for the lone-pair molecular configurations, when $P = (0, 1)$, $H(P = 0) = H(P = 1) = 0$, marking the alternative ion-pair configurations A^+B^- and A^-B^+, respectively, relative to the initial AO occupations $N^0 = (1, 1)$ in the assumed promolecular reference, in which both atoms contribute a single electron each to form the chemical bond.

The complementary descriptor $I^0(P) = 1 - H(P)$ of the bond IT ionicity (see Fig. 17), which determines the channel mutual information relative to the

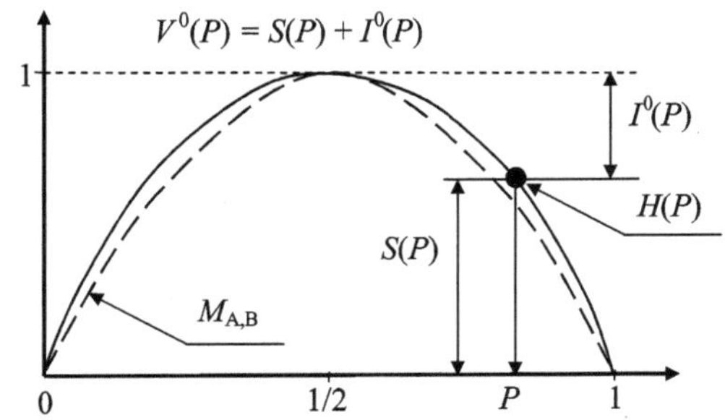

Fig. 17 Conservation of the overall entropic bond multiplicity $V^0(P) = 1$ bit in the 2-AO model of the chemical bond. It combines the conditional entropy (average noise, bond covalency) $S(P) = H(P)$ and the mutual information (information capacity, bond ionicity) $I^0(P) = 1 - H(P)$. In MO theory the direct bond order of Wiberg is represented by (*broken-line*) parabola $M_{A,B}(P) = 4P(1 - P) \equiv 4PQ$

promolecular input, reaches the highest value for the two limiting (electron-transfer) configurations: $I^0(P = 0) = I^0(P = 1) = H(½) = 1$ bit, and identically vanishes for the purely covalent, symmetric bond, $I^0(P = ½) = 0$. As explicitly shown in Fig. 17, these two components of the chemical bond multiplicity compete with one another, yielding the conserved overall IT bond index $V^0(P) = S(P) + I^0(P) = 1$ bit, marking a single bond in OCT in the whole range of admissible bond polarizations $P \in [0, 1]$. This simple model thus properly accounts for the competition between the bond covalency and ionicity, while preserving the single IT bond order measure reflected by the conserved overall multiplicity of the chemical bond.

Similar effects transpire from the quadratic bond indices formulated in the MO theory [81, 82, 87]. The corresponding plot of the Wiberg bond order for this model is given by the parabola (see Fig. 17):

$$M_{A,B}(P) = [\gamma_{A,B}(P)]^2 = 4PQ = 4P(1 - P), \tag{37}$$

which closely resembles the IT-covalent plot $S(P) = H(P)$ in the figure.

7 Localized Bonds in Diatomic Fragments

The bond components of Fig. 17 can be also decomposed into the corresponding atomic contributions [9]. In communication theory this partition is accomplished by using the partial (*row*) subchannels of Fig. 18, each determining communications originating from the specified atomic input. The partial entropy covalencies of the given atomic channel, calculated for the full (unit) probability of its input, again recover the binary entropy estimate of Fig. 17. However, the partial mutual information indices of the bond IT iconicity (charge transfer) have to be calculated for

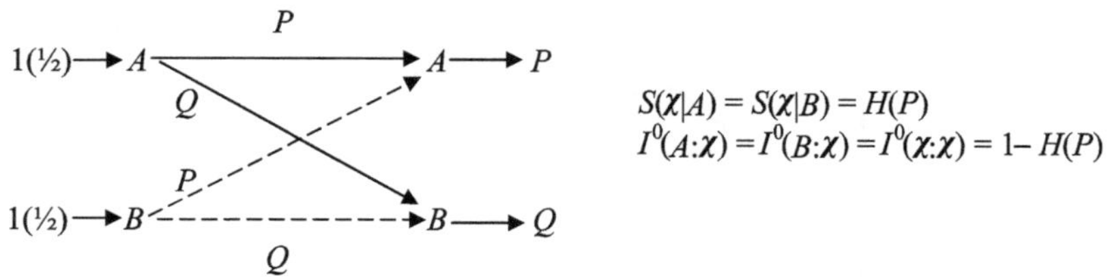

Fig. 18 The elementary partial (*row*) subchannels due to atomic inputs *A* (*solid lines*) and *B* (*broken lines*) in the 2-AO model of the chemical bond and their IT covalency/ionicity components

the full list of inputs, thus being equal to the promolecular mutual information index of Fig. 16.

In typical SCF LCAO MO calculations, the lone pairs of the *valence*- and/or *inner*-shell electrons can strongly affect the overall IT descriptors of the chemical bonds. Elimination of such *lone*-pair contributions in the resultant IT bond indices of diatomic fragments of molecules requires an *ensemble* (flexible input) approach [10, 11, 56]. In this generalized procedure the input probabilities are derived from the *joint* (bond) probabilities of two AO centered on different atoms, which reflect the actual *simultaneous* participation of the given pair of basis functions in chemical bonds. Such an approach effectively projects out the spurious contributions due to the *inner*- and *outer*-shell AO, which are excluded from mixing into the delocalized, bonding MO combinations. This probability-weighting procedure is capable of reproducing the Wiberg bond orders in diatomics, at the same time providing the IT-covalent/ionic resolution of these overall bond indices.

Let us illustrate this weighting procedure in the 2-AO model of the preceding section. In the bond-weighted approach, one uses the elementary subchannels of Fig. 18 and their partial entropy/information descriptors $\{S(\boldsymbol{\chi}|i)\}$, $I^0(i{:}\boldsymbol{\chi}) = I^0(\boldsymbol{\chi}{:}\boldsymbol{\chi})$; $\{i = A, B\}$, which are also listed in the diagram. In this particular case they are equal to the corresponding molecular conditional-entropy and mutual-information quantities of Fig. 16. Since these *row* descriptors represent the IT indices per electron in the diatomic fragment, these contributions have to be multiplied by $N_{AB} = N = 2$ in the corresponding resultant IT components and the overall measure of multiplicity of the effective diatomic bond. Using the off-diagonal joint probability $P(A \wedge B) = P(B \wedge A) = PQ = \gamma_{A,B}\gamma_{B,A}/4$ as the ensemble probability for both AO inputs gives the following average quantities for this model diatomic system (see Fig. 19):

$$S_{AB} = N[P(A \wedge B)S(\boldsymbol{\chi}|A) + P(B \wedge A)S(\boldsymbol{\chi}|B)] = 4PQH(P) = M_{A,B}H(P),$$
$$I^0_{AB} = N[P(A \wedge B)I^0(A{:}\boldsymbol{\chi}) + P(B \wedge A)I^0(B{:}\boldsymbol{\chi})] = 4PQ[1 - H(P)] = M_{A,B}[1 - H(P)],$$
$$V^0_{AB} = S_{AB} + I^0_{AB} = 4PQ = (\gamma_{A,B})^2 = M_{A,B}. \quad (38)$$

We have thus recovered the Wiberg index as the overall IT descriptor of the chemical bond in the 2-AO model, $V_{AB}{}^0 = M_{A,B}$, at the same time establishing

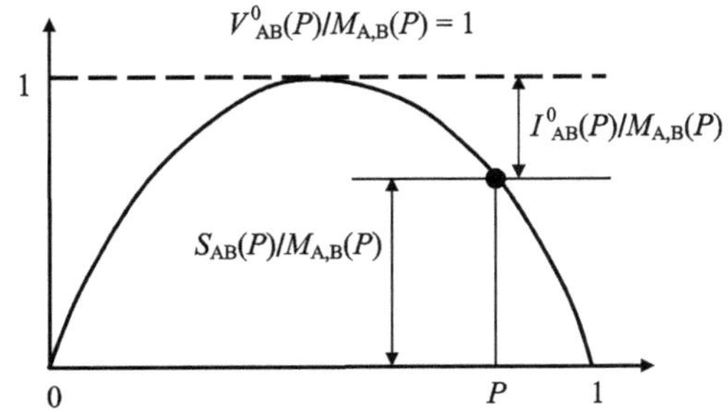

Fig. 19 Variations of the IT-covalent [$S_{AB}(P)$] and IT-ionic [$I_{AB}{}^0(P)$] bond components [in the $M_{A,B}(P)$ units] of the 2-AO model with changing MO polarization P and conservation of the relative total bond order $V_{AB}{}^0(P)/M_{A,B}(P) = 1$

Table 1 Comparison of the diatomic Wiberg indices and the entropy/information bond-multiplicity descriptors in selected molecules: RHF results in the minimum (STO-3G) basis set

Molecule	A–B	$M_{A,B}$	$V_{AB}{}^0$	S_{AB}	$I_{AB}{}^0$
H_2	H–H	1.000	1.000	1.000	0.000
F_2	F–F	1.000	1.000	0.947	0.053
HF	H–F	0.980	0.980	0.887	0.093
LiH	Li–H	1.000	1.000	0.997	0.003
LiF	Li–F	1.592	1.592	0.973	0.619
CO	C–O	2.605	2.605	2.094	0.511
H_2O	O–H	0.986	1.009	0.859	0.151
AlF_3	Al–F	1.071	1.093	0.781	0.311
CH_4	C–H	0.998	1.025	0.934	0.091
C_2H_6	C–C	1.023	1.069	0.998	0.071
	C–H	0.991	1.018	0.939	0.079
C_2H_4	C–C	2.028	2.086	1.999	0.087
	C–H	0.984	1.013	0.947	0.066
C_2H_2	C–C	3.003	3.063	2.980	0.062
	C–H	0.991	1.021	0.976	0.045
C_6H_6[a]	C_1–C_2	1.444	1.526	1.412	0.144
	C_1–C_3	0.000	0.000	0.000	0.000
	C_1–C_4	0.116	0.119	0.084	0.035

[a]For sequential numbering of carbon atoms in the ring

its covalent, S_{AB}, and ionic, $I_{AB}{}^0$, contributions. Again, these IT-covalency and IT-ionicity components compete with one another, while conserving the Wiberg bond order as the overall information measure of the bond multiplicity (in bits). This procedure can be generalized into the SCF MO calculations for general polyatomic systems and basis sets [10, 11, 56]. Illustrative RHF bond orders in diatomic fragments of representative molecules, for their equilibrium geometries in the minimum (STO-3G) basis set, are compared in Table 1. It follows from this numerical analysis that in polyatomics this weighting procedure gives rise to an excellent agreement with both the Wiberg bond orders and the accepted chemical intuition.

This ensemble averaging for the localized bond descriptors reproduces exactly the Wiberg bond order in diatomic molecules [56]. In a series of related compounds, e.g., in hydrides or halides, the trends exhibited by the entropic covalent and ionic components of a roughly conserved overall bond order also agree with intuitive expectations. For example, the single chemical bond between two "hard" atoms in HF appears predominantly covalent, while a substantial ionicity is detected for LiF, for which both Wiberg and information-theoretic results predicts roughly 3/2 bond, consisting of approximately 1 covalent and ½ ionic contributions. One also observes that all carbon–carbon interactions in the benzene ring are properly differentiated. The chemical orders of the multiple bonds in ethane, ethylene, and acetylene are properly reproduced and the triple bond in CO is correctly accounted for. As intuitively expected, the C–H bonds are seen to slightly increase their information ionicity when the number of these terminal bonds increases in a series: acetylene, ethylene, ethane.

8 Through-Space and Through-Bridge Bond Components

In OCT the direct communication $\chi_i \rightarrow \chi_j$ between the given pair of AO, manifested by the nonvanishing conditional probability $P(j|i) > 0$, reflects a "dialogue" between these basis functions in the molecular ground state. It indicates the existence of the *direct* (through-space) chemical bonding between these orbitals, due to their nonvanishing overlap/interaction giving rise to their constructive interference in the bonding MO combination. As we have demonstrated in the preceding section, the Wiberg bond-order contribution [see also Eqs. (20) and (22)],

$$\begin{aligned} M_{i,j} &= \gamma_{i,j}\gamma_{j,i} = 4d_{i,j}d_{j,i} \\ &= 4\langle j|\hat{\mathrm{P}}_\varphi|i\rangle\langle i|\hat{\mathrm{P}}_\varphi|j\rangle = 4(\langle j|\hat{\mathrm{P}}_\varphi|)\left[(|\hat{\mathrm{P}}_\varphi|i\rangle)(\langle i|\hat{\mathrm{P}}_\varphi|)\right](|\hat{\mathrm{P}}_\varphi|j\rangle) \equiv 4\left\langle j^b\middle|\hat{\mathrm{P}}_i^b\middle|j^b\right\rangle \\ &= 4\left|\langle i^b|j^b\rangle^2\right| \equiv 4\left|d_{i,j^b}\right|^2, \end{aligned} \tag{39}$$

provides a useful measure of the multiplicity of this *explicit* interaction. In this equation we have used the idempotency property of the projector $\hat{\mathrm{P}}_\varphi$ onto the bonding (doubly occupied) subspace of MO, $\hat{\mathrm{P}}_\varphi^2 = \hat{\mathrm{P}}_\varphi$. This operator determines the associated nonorthogonal projections of AO:

$$|\boldsymbol{\chi}^b\rangle = \hat{\mathrm{P}}_\varphi|\boldsymbol{\chi}\rangle = \{|\chi_i{}^b\rangle\}.$$

The direct "bond order" measures the magnitude of the overlap $d_{i,j}$ between bond projections of the interacting basis functions.

However, communications between χ_i and χ_j can be also realized indirectly via a cascade involving other orbitals, i.e., as a "gossip" spread through the remaining AO, $\boldsymbol{\chi}' = \{\chi_{k\neq(i,j)}\}$. For example, this *implicit* scattering process may involve a single AO intermediates, in the *single*-step "cascade" propagation: $\chi_i \rightarrow \boldsymbol{\chi}' \rightarrow \chi_j$. However, since in the molecular channel each AO both emits and receives the signal to/from remaining basis functions, this process may also involve any admissible *multi*step cascade, $\chi_i \rightarrow \{\boldsymbol{\chi}' \rightarrow \boldsymbol{\chi}' \rightarrow \ldots . \rightarrow \boldsymbol{\chi}'\} \rightarrow \chi_j$, in which the consecutive sets of AO intermediaries $\{\boldsymbol{\chi}' \rightarrow \boldsymbol{\chi}' \rightarrow \ldots . \rightarrow \boldsymbol{\chi}'\}$ form an effective multistep "bridge" for the information scattering. The OCT formalism has been recently extended to tackle such *indirect* chemical communications [59–62]. The corresponding Wiberg-type bond multiplicities and the associated IT bond descriptors for such cascade information channels have been proposed, capable of describing these through-bridge chemical interactions.

In MO theory the chemical coupling between, say, two (valence) AO or general basis functions originating from different atoms is strongly influenced by their direct overlap and interaction. Together these two factors condition the bonding effect experienced by electrons occupying their bonding combination in the molecule, compared to the *non*bonding reference of electrons on the separated AO. This through-*space* bonding mechanism is generally associated with an accumulation of the valence electrons in the region between the two nuclei, due to the constructive interference in the bonding MO. Indeed, such "shared" bond charge is customarily regarded as the prerequisite for the bond *covalency* in the direct interaction between the two AO. It is also reflected by the familiar *covalent* Valence-Bond (VB) structure. In Sects. 4 and 5 a similar effect of the bonding accumulation of the information densities relative to the promolecular distribution has been detected. The complementary, *ionicity* aspect of the direct chemical bonding is manifested by the MO polarization, reflecting the charge transfer effects, or by the participation of the orthogonal part of the *ionic* structure in the ground-state wave function in VB theory. In OCT the bond ionicity descriptor reflects a degree of "localization" (determinicity) in direct communications between AO, while the complementary bond covalency index measures the "delocalization" (noise) aspect of the direct orbital channel.

To summarize, the *direct* ("through-space") bonding interaction between neighboring atoms is in general associated with the presence of the bond charge between the two nuclei. However, for more distant atomic partners such an accumulation of valence electrons can be absent, e.g., in the cross-ring π-interactions in benzene or between the bridgehead carbon atoms in small propellanes [59]. Nonetheless, the bonding interaction lacking this accumulation of the charge (information) can be still realized *indirectly*, through the neighboring AO intermediaries forming a "bridge" for an effective interaction between more distant (terminal) AO, e.g., in the cross-ring interactions between two *meta*- or *para*-carbons in benzene, two bridgehead carbons in small propellanes [59], or higher order neighbors in the polymer chain [61, 62]. This indirect mechanism was shown to reflect the implicit dependencies between the AO bond projections $\boldsymbol{\chi}^b$ [60], which reflect the resultant AO participations in all chemical bonds in the molecular system under

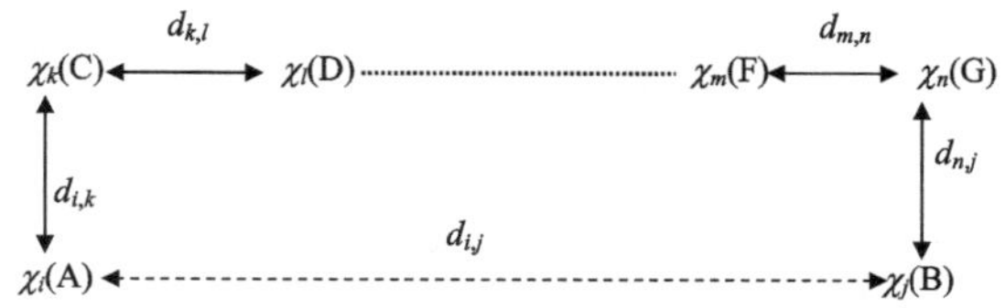

Fig. 20 Direct (through-space) chemical interaction (*broken line*) between orbitals χ_i and χ_j contributed by atoms A and B, respectively, and the indirect (through-*bridge*) interaction (*solid/pointed lines*), through *t* AO-intermediaries $(\chi_k, \chi_l, \ldots, \chi_m, \chi_n) = \{\chi_r, r = 1, 2, \ldots, t\}$ contributed by the neighboring bonded atoms (C, D, ..., F, G), respectively. The strength of each partial (direct) interaction is reflected by the magnitude of the corresponding elements of the density matrix $\mathbf{d} = \langle \boldsymbol{\chi}^b | \boldsymbol{\chi}^b \rangle = \{d_{i,j}\}$, which measure the overlaps between the bond projections $\boldsymbol{\chi}^b$ of AO

consideration. Indeed, the orthonormality constraints imposed on the occupied MO imply the *implicit* dependencies between the (nonorthogonal) bond projections of AO on different atoms, due to real chemical bridges in molecules (see Fig. 20).

In this generalized outlook on the chemical bond-index concept, emerging from both the Wiberg, quadratic measure of MO theory and the IT bond multiplicities of OCT, one thus identifies the chemical bond "order" as a measure of a "*dependence*" (*non*additivity) between orbitals on different atomic centers. On one hand, this dependence can be realized *directly* (through space), by the constructive interference of orbitals (probability amplitudes) on two atoms, which generally increases the electron density between them. On the other hand, it can also have an *indirect* origin, through the dependence on orbitals of the remaining AIM used to construct the occupied MO subspace $\boldsymbol{\varphi}^{\mathrm{o}} = \boldsymbol{\chi}\mathbf{C}^{\mathrm{o}}$. These implicit ("geometrical") dependencies are embodied in the (idempotent) density matrix:

$$\mathbf{d} = \langle \boldsymbol{\chi}^b | \boldsymbol{\chi}^b \rangle = \boldsymbol{\gamma}/2 = \mathbf{C}^o \mathbf{C}^{o\dagger} = \{d_{i,j} = \langle i^b | j^b \rangle\}, \quad \mathbf{d}^n = \mathbf{d}. \tag{40}$$

Each pair of AO on different atoms is thus capable of exhibiting a partial through-*space* and through-*bridge* bond components. The order of the former quickly vanishes with an increase of the *inter*-atomic separation, and when the interacting AO are heavily engaged in forming chemical bonds with other atoms. At these separations the indirect bond orders can still assume appreciable values, when the remaining atoms form an effective bridge of the neighboring, chemically interacting atoms, which links the specified pair of terminal AO. The bridging atoms must be mutually bonded to generate the appreciable through bridge overlap of the interacting AO, so that a variety of significant bridges is practically limited to real chemical bridges of atoms in the molecular structural formula.

The representative indirect bond overlap through the *t*-bridge of Fig. 20, $S_{i,j}(k, l, .., m, n) = S_{i,j}(\{r\})$, constitutes a natural generalization of its direct, through-space analog by additionally including the product of *bond* projectors onto the indicated intermediate AO,

$$S_{i,j}(k,l,\ldots m,n) = \left\langle i^b \middle| \prod_{r=1}^{t} \hat{\mathrm{P}}_r^b \middle| j^b \right\rangle \equiv \left\langle i^b \middle| \hat{\mathrm{P}}_{t-\mathrm{bridge}}^b \middle| j^b \right\rangle. \tag{41}$$

For specific *single-* or *multi-*step bridges realized by the indicated AO intermediates, this indirect bond overlap is given by the relevant products of direct overlaps in the bridge:

$$S_{i,j}(k) = d_{i,k}d_{k,j}, S_{i,j}(k,l) = d_{i,k}d_{k,l}d_{l,j}, \text{etc.}$$

Hence, for a general t-bridge of Fig. 20 one finds

$$S_{i,j}(k,l,..,m,n) = d_{i,k}d_{k,l}\ldots d_{m,n}d_{n,j}. \tag{42}$$

The square of this generalized bond overlap then defines the associated Wiberg-type bond order for such implicit interaction between orbitals χ_i and χ_j via the specified t-bridge:

$$\begin{aligned} M_{i,j}(k,l,\ldots,m,n) &= 2^{2t}|S_{i,j}(k,l,..,m,n)|^2 \\ &= \gamma_{i,k}\{\gamma_{k,l}\cdots[\gamma_{m,n}(\gamma_{n,j}\gamma_{j,n})\gamma_{n,m}]\cdots\gamma_{l,k}\}\gamma_{k,i} \\ &= M_{i,k}M_{k,l}M_{l,m}\ldots M_{m,n}M_{n,j}. \end{aligned} \tag{43}$$

The sum of contributions due to the most important (chemical) bridges $\{\alpha\}$:

$$M_{i,j}(\text{bridge}) = \sum_{\alpha} M_{i,j}(\alpha), \tag{44}$$

then determines the overall indirect Wiberg-type bond order, which supplements the direct component $M_{i,j}$ in the full quadratic bond-multiplicitiy index, between terminal orbitals χ_i and χ_j in presence of remaining AO:

$$M(i,j) = M_{i,j} + M_{i,j}(\text{bridge}), \tag{45}$$

As an illustration let us examine the indirect π-bonds between carbon atoms in benzene and butadiene, in the familiar Hückel approximation [59]. For the consecutive numbering of carbons in the ring/chain, the relevant CBO matrix elements in benzene read

$$\gamma_{i,i} = 1, \gamma_{i,i+1} = 2/3, \gamma_{i,i+2} = 0, \quad \gamma_{i,i+3} = -1/3,$$

while the *off*-diagonal part of the CBO matrix in butadiene is fully characterized by the following elements:

$$\gamma_{1,2} = \gamma_{3,4} = 2/\sqrt{5}, \quad \gamma_{1,3} = \gamma_{2,4} = 0, \gamma_{1,4} = -1/\sqrt{5}, \quad \gamma_{2,3} = 1/\sqrt{5}.$$

They generate the associated direct (through-space) bond multiplicities:

$$\text{benzene}: M_{i,i+1} = 0.44(\text{ortho}), \quad M_{i,i+2} = 0(\text{meta}), \quad M_{i,i+3} = 0.11(\text{para});$$
$$\text{butadiene}: M_{1,2} = M_{3,4} = 0.80, \quad M_{1,3} = M_{2,4} = 0, \quad M_{1,4} = M_{2,3} = 0.20.$$

These direct bond orders are complemented by the following estimates of the resultant multiplicities of the indirect π-interactions due to chemical bridges:

$$\text{benzene}: M_{i,i+1}(\text{bridge}) = 0.06(\text{ortho}), \quad M_{i,i+2}(\text{bridge}) = 0.30(\text{meta}),$$
$$M_{i,i+3}(\text{bridge}) = 0.18(\text{para});$$
$$\text{butadiene}: M_{1,2}(\text{bridge}) = M_{3,4}(\text{bridge}) = 0.03, \ M_{1,3}(\text{bridge}) = M_{2,4}(\text{bridge}) = 0.32,$$
$$M_{1,4}(\text{bridge}) = M_{2,3}(\text{bridge}) = 0.13.$$

Together these contributions give rise to the following resultant bond orders of Eq. (45):

$$\text{benzene}: M(\text{para}) \cong M(\text{meta}) = 0.3 < M(\text{ortho}) = 0.5.$$
$$\text{butadiene}: M(1-2) = 0.83 > M(1-4) = 0.33 \cong M(1-3) = 0.32.$$

Of interest also is a comparison of the bond-order contributions in benzene realized through the ring bridges of increasing length:

$$M_{i,i+2}(i+1) = (M_{i,i+1})^2 = 0.20,$$
$$M_{i,i+3}(i+1, i+2) = (M_{i,i+1})^3 = 0.09,$$
$$M_{i,i+4}(i+1, i+2, i+3) = (M_{i,i+1})^4 = 0.04,$$
$$M_{i,i+5}(i+1, i+2i+3, i+4) = (M_{i,i+1})^5 = 0.02.$$

Thus, the longer the bridge, the smaller indirect bond order it contributes. The model and Hartree-Fock calculations for representative polymers [61, 62] indicate that the range of bridge interactions is effectively extended to *third*-order neighbors in the chain, for which the direct interactions practically vanish.

The artificial distinction in Wiberg's multiplicity scheme of the π-interactions with the vanishing direct CBO matrix element as nonbonding is thus effectively removed, when the through-bride contributions are also taken into account. One observes the differences in composition of the resultant indices for the *cross*-ring interactions in benzene: the *para* interactions exhibit comparable through-space and through-bridge components, the *meta* multiplicities are realized exclusively through bridges, while the strongest *ortho* bonds have practically direct, through-space origin. A similar pattern can be also observed in butadiene.

The conditional probabilities of Eq. (26) determine the molecular information channel for the mutual communications between AO, which generate the associated covalency (noise) and ionicity (information flow) descriptors of the direct chemical bonds. One similarly derives the corresponding entropy/information multiplicities for the indirect interactions between the specified (terminal) orbitals χ_i and χ_j, generated from descriptors of the associated AO information cascades for the most important (chemical) bridges [11, 59, 61]. The resulting IT indices of specific bridge interactions have been shown to compare favorably with the generalized Wiberg measures of Eq. (43). For an attempt to separate the direct and indirect energy contributions, the reader is referred to [91].

9 Information Probes of Elementary Reaction Mechanisms

Interesting new results in the IT studies of the elementary reaction mechanisms have recently been obtained in the Granada group [92–94]. Both the *global* (Shannon) and *local* (Fisher) information measures have been used in these investigations. The course of two representative reactions has been examined: of the *radical abstraction* of hydrogen (*two*-step mechanism),

$$H^{\bullet} + H_2 \rightarrow H_2 + H^{\bullet}, \tag{46}$$

which requires extra energy to proceed, and of the *nuclephilic substitution* in the hydride exchange (S_N2, *one*-step mechanism):

$$H^{-} + CH_4 \rightarrow CH_4 + H^{-}. \tag{47}$$

The abstraction process proceeds by homolysis and is kinetically of the *first*-order (S_N1-like). It involves two steps: formation of new radicals, via the homolytic cleavage of the nonpolar, perfectly covalent bond in H_2, in absence of any electrophile or nucleophile to initiate the heterolytic pattern, and the subsequent recombination of a new radical with another radical species. The hydride exchange is an example of the kinetically *second*-order, the *first*-order in both the *incoming* (nucleophile) and *leaving* (nucleofuge) hydride groups. It proceeds via the familiar Walden-inversion *Transition State* (TS) in a single, concerted reaction.

The central quantities of these IT studies are the Shannon entropies in both the position ($\boldsymbol{r}$) and momentum ($\boldsymbol{p}$) spaces,

$$S_r = -\int \rho(\boldsymbol{r}) \ln \rho(\boldsymbol{r}) d\boldsymbol{r}, \qquad S_p = -\int \eta(\boldsymbol{p}) \ln \eta(\boldsymbol{p}) d\boldsymbol{p}, \tag{48}$$

and the related Fisher information measures:

$$I_r = \int |\nabla\rho(\boldsymbol{r})|^2/\rho(\boldsymbol{r}) d\boldsymbol{r}, \qquad I_p = \int |\nabla\eta(\boldsymbol{p})|^2/\eta(\boldsymbol{p}) d\boldsymbol{p}. \tag{49}$$

Here, $\eta(\boldsymbol{p})$ stands for the momentum density of electrons, efficiently generated from the Fourier transforms of the known position-space MO.

The reaction profiles of these information probes have discovered the presence of additional features of the two reaction mechanisms by revealing the chemically important regions where the bond forming and bond breaking actually occur. These additional features cannot be directly identified from the energy profile alone and from the structure of the TS densities involved. Consistency of predictions resulting from the global (Shannon) [92] and local (Fisher) [93, 94] information measures has additionally confirmed a more universal and unbiased character of these findings.

Indeed, either of the two complementary Shannon entropies for the model radical abstraction reaction displays a richer structure than the associated energy profile, which only exhibits one maximum at the TS point along the reaction coordinate. The position entropy $S_{\boldsymbol{r}}$ displays a local maximum at this TS structure and two minima in its close proximity, whereas the momentum entropy $S_{\boldsymbol{p}}$ exhibits the global minimum at TS complex and two maxima at points slightly more distant from TS than the corresponding positions of the $S_{\boldsymbol{r}}$ minima. It thus follows from these entropy curves that the approach of the hydrogen molecule by the incoming hydrogen in the proximity of TS first localizes ρ in preparation for the bond rupture, which also implies an associated increase in the kinetic energy (delocalization of η). This preparatory stage is identified by the local minima of $S_{\boldsymbol{r}}$ (maxima of $S_{\boldsymbol{p}}$). Next, when the system relaxes and the new bond is formed at TS, the position (momentum) densities become more delocalized (localized). This is indeed manifested by the corresponding maximum (minimum) features of $S_{\boldsymbol{r}}$ ($S_{\boldsymbol{p}}$). The bond-breaking process requires energy, as indeed witnessed by an earlier maximum of $S_{\boldsymbol{p}}$ in the entrance valley of the reaction *Potential Energy Surface* (PES), which is subsequently dissipated by relaxing the structure at TS. In other words, the reaction complex first gains the energy required for the bond dissociation, and then the position-space density gets localized to facilitate the bond cleavage, which in turn induces the energy/density relaxation towards the TS structure.

Therefore, the entropy representation of the reaction mechanism reveals the whole complexity of this transformation, while the associated *Minimum Energy Path* (MEP) profile only localizes the transition state on PES, missing the crucial transitory localization/delocalization and relaxational phenomena involved in this *two*-step process.

The corresponding $S_{\boldsymbol{r}}(S_{\boldsymbol{p}})$ plot for the hydride-exchange S_N2 process again exhibits the maximum (minimum) at TS, with two additional minima (maxima) in its vicinity, where the bond breaking is supposed to occur. These additional, *pre*- and *post*-TS features are symmetrically placed in the entrance and exit valleys, relative to TS structure, but now they appear at roughly the same values of the *Intrinsic Reaction Coordinate* (IRC) in both the position and momentum representations. This is in contrast to the *two*-stage abstraction mechanism, where in the *entrance* valley the ***p***-space maximum of the Shannon entropy has preceded the associated minimum observed in the ***r***-space. This simultaneous ***r***-localization (***p***-delocalization) may be indicative of the *single*-step mechanism in which the approach of the nucleophile is perfectly synchronized with the concomitant departure of the nucleofuge, so that the bond forming and bond breaking occur in a concerted manner.

It should be observed that both these displacements increase the system energy. First, as the nucleophile approaches, this energy is required to overcome the repulsion between reactants and create the *position* localization (*momentum* delocalization) facilitating the bond weakening. As the reaction progresses forward, the energy continues to grow towards the maximum at TS, when the sufficient threshold of the new chemical bond has already been reached to start the structure relaxation inducing the reverse $\boldsymbol{r}$-delocalization ($\boldsymbol{p}$-localization) processes leading to the S_r (S_p) maximum (minimum) at TS. This synchronous transformation picture is indeed customarily associated with this particular reaction.

The same sequences of the chemical events are seen in the complementary Fisher-information analysis of the structural features of distributions in both spaces. For the hydrogen abstraction reaction one observes with the progress of IRC towards TS that, relative to the separated reactants reference, both I_r and I_p at first decrease their values, thus marking a lower average gradient content of the associated probability amplitudes (wave functions) in both spaces, i.e., a more regular/uniform distribution (less structure, "order"). The I_p profile is seen to exhibit a faster decay towards the local minimum preceding TS, where I_p reaches the maximum value. These more uniform momentum densities also correspond to the local maxima of the system chemical hardness, with TS marking the local minimum of the latter. The I_r monotonically decreases towards the minimum value at TS, thus missing the additional extrema observed in the S_r plot, which have been previously associated with the bond homolysis. Thus, the I_r is not capable of describing the bond breaking/forming processes, which is clearly uncovered by I_p.

The disconcerted manner of the elementary bond forming and bond breaking is directly seen in the corresponding bond-length plots: the breaking of the bond occurs first, and then the system stabilizes by forming the TS structure. Additional insight into the density reconstruction in this homolytic bond rupture comes from examining the corresponding plots of the system dipole moment, reflecting the charge distortion during reaction progress. The observed behavior of these functions is opposite to I_p. Therefore, in regions of the minimum I_p the dipole moment reaches the maximum value and vice versa. As intuitively expected, the dipole moment identically vanishes at both TS and in separated reactants/products.

Finally, we turn to the Fisher-information analysis of the hydride-exchange reaction involving the heterolytic bond cleavage, with an accompanying exchange of charge between reactants. The corresponding I_r and I_p functions of the reaction coordinate now display similar behavior, both exhibiting maxima at TS, where the Shannon entropies have indicated a more delocalized position density and relatively localized momentum density. One also observes two minima of I_r and I_p in the proximity of TS. These IRC values coincide with the bond breaking/forming regions, and the change in the curvature of the bond-elongation curves for the incoming and outgoing nucleophile marks the start of an increase in the gradient content of the momentum density, towards the maximum value at TS structure.

10 Conclusion

Information theory has been shown to provide a novel and attractive perspective on the entropic origins of the chemical bond. It also offers a complementary outlook on the transformation of the electronic information content in the elementary chemical reactions. In this short overview, we have first introduced the key IT concepts and techniques to be used in such a complementary analysis of electron distributions in molecular systems. They have been subsequently applied to explore the bonding pattern in typical molecules: in terms of the information distribution, the bond localization/multiplicity, and its ionic/covalent composition. The use of the information densities as local probes of electronic distributions in molecules has been advocated and the importance of the nonadditive entropy/information measures in extracting subtle changes due to the bond formation has been stressed. The use of the CG density, of the nonadditive Fisher information (electronic kinetic energy) in the AO resolution, as an efficient localization probe of the direct chemical bonds has been validated.

The OCT, in which molecules are regarded as information systems, has been shown to be capable of tackling several classical issues in the theory of bond multiplicities and their covalent/ionic components. The *off*-diagonal (inter-orbital) conditional probabilities it generates are proportional to the quadratic bond indices formulated in MO theory, and hence the strong *inter*-orbital communications correspond to strong Wiberg bond orders. The IT "fingerprints" of typical elementary mechanisms of chemical reactions have also revealed new stages in the bond-breaking–bond-forming processes involved. These illustrative results demonstrate that the alternative information probes increase our understanding of the complex chemical bond phenomena, with the communication-*noise* (covalency) and information-*flow* (ionic) measures of the IT bond orders adequately reflecting the accepted chemical intuition. This development extends our understanding of the chemical bond from the complementary IT viewpoint.

The CG criterion explores the nonadditive part of the Fisher information in AO resolution, which reflects the electron *delocalization* phenomena due to the direct chemical bonds. It has been also demonstrated elsewhere [29] that the key conditional probability ingredient of the original ELF concept [30–32] is also related to the nonadditive Fisher information in MO resolution. The inverse of the latter thus quantifies a degree of a *localization* of the system electrons in a molecule, in accordance with an earlier Fermi-hole analysis by Luken et al. [95, 96]. The spatial organization of the electronic structure can be also probed stochastically, by combining the path integral Markovian pair-conditional probability density with the basic concepts of the catastrophe theory [97], in a way consistent with the Heisenberg and Pauli principles of quantum mechanics. This analysis provides the Markovian derivation of the original ELF and extends this localization concept to new Markovian-type ELF classes.

The entropy/information descriptors can be generated for both the molecule as a whole and its localized fragments. The relevant OCT treatment of diatomic subsystems has been described. This approach reproduces the Wiberg bond multiplicity in diatomic molecules and closely approximates this index in polyatomic molecules. The extra computational effort of such an IT analysis of the molecular bonding patterns is negligible, compared to the cost of standard SCF LCAO MO calculations of the molecular electronic structure, since practically all computations using orbital approximation already determine the CBO and kinetic energy data required in the OCT probe of IT bond multiplicities and the CG localization of direct chemical bonds.

It has also been stressed that chemical interactions between the specified pair of AO have both the direct (through-space) and, hitherto neglected, indirect (through-bridge) components. The former reflects the direct interference between bonded atoms, while the latter is realized indirectly, through the remaining AIM which form effective bridges for the chemical coupling between more distant atoms. The most efficient bridges are the real (chemical) bridges of bonded atoms connecting such "terminal" atoms in question. Therefore, the bonded status of the given pair of atoms can be felt also at larger separations provided there exist real bridge of the direct chemical bonds connecting them. This implicit coupling of basis functions in molecular states has been first conjectured to explain the central bond in small propellanes, lacking the charge accumulation between the bridgehead carbons. This difficulty has also prompted the alternative proposition of the VB-inspired *charge-shift* [98] mechanism. The latter attributes this indirect bond to the instantaneous charge fluctuations between the bridgehead carbons.

We have also discussed the indirect Wiberg-type indices for this "through-*bridge*" interaction in molecular systems, which complement the familiar Wiberg bond orders of the "through-*space*" bonding. The bridge interactions have been shown to have important implications for the resultant multiplicities of π-interactions between more distant carbon atoms in hydrocarbons. For example, in benzene the *ortho*-carbons exhibit a strong Wiberg bond multiplicity, of almost exclusively through-space origin, while the *cross*-ring interactions between the *meta*- and *para*-carbons were shown to be described by much smaller, but practically equalized overall bond orders. The latter can be distinguished by their direct/indirect composition: the *meta* bonds have been shown to be realized exclusively through bridges, while the *para* bonds exhibit comparable direct and indirect contributions.

In OCT the "*explicit*" (through-space) bond component contributed by the specified pair of interacting AO originates from their mutual probability scattering. Its covalency is thus generated by finite conditional probabilities of their direct communications in the molecule, related to the square of the corresponding element of the system density matrix and hence also to the associated Wiberg index. These direct AO communications are thus determined by the electron delocalization throughout the system occupied (bonding) subspace of MO. The "*implicit*" (through-bridge) bond contributions can be similarly viewed as resulting from the indirect (cascade) information propagation via the bridging AO. In a sense, the through-space bonding can be regarded as reflecting the direct "conversation" between AO, while the through-bridge channel(s) can be compared to a chatty talk reporting "hearsay," the "*rumor*" spread between the terminal AO via the connecting chain of the AO intermediaries of the bridge under consideration. Together the direct ("dialogue") and indirect ("gossip") contributions to the chemical bond order determine the resultant bond multiplicity between the specified AO in molecular environment. Each pair of atoms thus exhibits partial through-*space* and through-*bridge* bond components. The "order" of the former quickly vanishes with the increasing *inter*-atomic separation. The overall bond order between more distant atomic partners can still assume appreciable values, when some of the remaining atoms form an effective bridge of the neighboring, chemically bonded atoms, which links the two AIMs in question.

Therefore, the requirement for a *non*vanishing density-matrix element coupling the two AO in the molecule, which in MO theory reflects their directly bonding status, is not essential for an existence of their effective through-bridge interaction, provided the two AO chemically couple to the bonded chain of orbitals connecting them.

This novel perspective on the entropic origins of chemical bonds is very much in the spirit of the Eugene Wigner's observation, often quoted by Walter Kohn, that the understanding in science requires insights from several different points of view. The IT probes of molecular systems and chemical reactions generate such an additional perspective on the genesis of chemical bonds and the course of elementary reaction mechanisms. It complements the familiar MO interpretations of quantum chemistry and gives rise to a deeper understanding of these complex phenomena.

References

1. Fisher RA (1925) Proc Cambridge Philos Soc 22:700
2. Frieden BR (2004) Physics from the Fisher information—a unification, 2nd edn. Cambridge University, Cambridge
3. Shannon CE (1948) Bell System Tech J 27:379, 623
4. Shannon CE, Weaver W (1949) The mathematical theory of communication. University of Illinois, Urbana
5. Kullback S, Leibler RA (1951) Ann Math Stat 22: 79
6. Kullback S (1959) Information theory and statistics. Wiley, New York
7. Abramson N (1963) Information theory and coding. McGraw-Hill, New York
8. Pfeifer PE (1978) Concepts of probability theory, 2nd edn. Dover, New York
9. Nalewajski RF (2006) Information theory of molecular systems. Elsevier, Amsterdam
10. Nalewajski RF (2010) Information origins of the chemical bond. Nova, New York
11. Nalewajski RF (2012) Perspectives in electronic structure theory. Springer, Heidelberg (in press)
12. Nalewajski RF (2011) In: Hong WI (ed) Mathematical chemistry. Nova, New York, p 247
13. Nalewajski RF (2011) In: Putz MV (ed) Chemical information and computation challenges in 21st century. Nova Science Publishers, New York (in press)
14. Nalewajski RF, deSilva P, Mrozek J (2011) In: Roy AK (ed) Theoretical and computational developments in modern density functional theory. Nova Science Publishers, New York (in press)

15. Nalewajski RF, Świtka E, Michalak A (2002) Int J Quantum Chem 87:198
16. Nalewajski RF, Broniatowska E (2003) J Phys Chem A 107:6270
17. Nalewajski RF, Parr RG (2000) Proc Natl Acad Sci USA 97:8879
18. Nalewajski RF, Loska R (2001) Theoret Chem Acc 105:374
19. Nalewajski RF (2002) Phys Chem Chem Phys 4:1710
20. Nalewajski RF (2003) Chem Phys Lett 372:28
21. Parr RG, Ayers PW, Nalewajski RF (2005) J Phys Chem A 109:3957
22. Nalewajski RF (2003) Adv Quant Chem 43:119
23. Nalewajski RF, Broniatowska E (2007) Theoret Chem Acc 117:7
24. Nalewajski RF, Parr RG (2001) J Phys Chem A 105:7391
25. Hirshfeld FL (1977) Theoret Chim Acta (Berl) 44:129
26. Nalewajski RF (2008) Int J Quantum Chem 108:2230
27. Nalewajski RF (2010) J Math Chem 47:667
28. Nalewajski RF, deSilva P, Mrozek J (2010) J Mol Struct Theochem 954:57
29. Nalewajski RF, Köster AM, Escalante S (2005) J Phys Chem A 109:10038
30. Becke AD, Edgecombe KE (1990) J Chem Phys 92:5397
31. Silvi B, Savin A (1994) Nature 371:683
32. Savin A, Nesper R, Wengert S, Fässler TF (1997) Angew Chem Int Ed Engl 36:1808
33. Nalewajski RF (2000) J Phys Chem A 104:11940
34. Nalewajski RF, Jug K (2002) In: Sen KD (ed) Reviews of modern quantum chemistry: a celebration of the contributions of Robert G. Parr, Vol. I. World Scientific, Singapore, p. 148
35. Nalewajski RF (2004) Struct Chem 15:391
36. Nalewajski RF (2004) Mol Phys 102:531, 547
37. Nalewajski RF (2005) Theoret Chem Acc 114:4
38. Nalewajski RF (2008) J Math Chem 43:265
39. Nalewajski RF (2009) Int J Quantum Chem 109:425
40. Nalewajski RF (2005) Mol Phys 103:451
41. Nalewajski RF (2006) Mol Phys 104:365, 493, 1977, 2533
42. Nalewajski RF (2005) J Math Chem 38:43
43. Nalewajski RF (2008) J Math Chem 44:414
44. Nalewajski RF (2009) J Math Chem 45:607
45. Nalewajski RF (2009) J Math Chem 45:709
46. Nalewajski RF (2009) J Math Chem 45:776
47. Nalewajski RF (2009) J Math Chem 45:1041
48. Nalewajski RF (2010) J Math Chem 47:709
49. Nalewajski RF (2008) J Math Chem 43:780
50. Nalewajski RF (2007) J Phys Chem A 111:4855
51. Nalewajski RF (2006) Mol Phys 104:3339
52. Nalewajski RF (2003) J Phys Chem A 107:3792
53. Nalewajski RF (2006) Mol Phys 104:255
54. Nalewajski RF (2004) Ann Phys (Leipzig) 13:201
55. Nalewajski RF (2009) Adv Quant Chem 56:217
56. Nalewajski RF, Szczepanik D, Mrozek J (2011) Adv Quant Chem 61:1
57. Nalewajski RF (2010) J Math Chem 47:692, 709, 808
58. Nalewajski RF (2011) 49:592
59. Nalewajski RF (2011) J Math Chem 49:371, 546, 806
60. Nalewajski RF, Gurdek P (2011) J Math Chem 49:1226
61. Nalewajski RF (2012) Int J Quantum Chem 112:2355
62. Nalewajski RF, Gurdek P (2011) Struct Chem (M. Witko issue) 23:1383
63. Nalewajski RF (2011) J Math Chem 49:2308
64. vonWeizsäcker CF (1935) Z Phys 96:431
65. Kohn W, Sham LJ (1965) Phys Rev 140A:1133
66. Hohenberg P, Kohn W (1964) Phys Rev 136B:864

67. Nalewajski RF, Świtka E (2002) Phys Chem Chem Phys 4:4952
68. Capitani JF, Nalewajski RF, Parr RG (1982) J Chem Phys 76:568
69. Cortona P (1991) Phys Rev B 44:8454
70. Wesołowski T, Warshel A (1993) J Phys Chem 97:8050
71. Wesołowski T (2004) J Am Chem Soc 126:11444
72. Wesołowski T (2004) Chimia 58:311
73. Gordon RG, Kim YS (1972) J Chem Phys 56:3122
74. Feinberg MJ, Ruedenberg K, Mehler EL (1970) Adv Quant Chem 5:28
75. Goddard WA, Wilson CW (1972) Theoret Chim Acta (Berl) 26:195, 211
76. Feinberg MJ, Ruedenberg K (1971) J Chem Phys 54:1495
77. Feinberg MJ, Ruedenberg K (1971) J Chem Phys 55:5804
78. Nalewajski RF (2009) Int J Quantum Chem 109:245
79. Dirac PAM (1958) The principles of quantum mechanics, 4th edn. Clarendon, Oxford
80. Wiberg KA (1968) Tetrahedron 24:1083
81. Nalewajski RF, Köster AM, Jug K (1993) Theoret Chim Acta (Berl) 85:463
82. Nalewajski RF, Mrozek J (1994) Int J Quantum Chem 51:187
83. Nalewajski RF, Mrozek J, Mazur G (1996) Can J Chem 100:1121
84. Nalewajski RF, Mrozek J, Michalak A (1997) Int J Quantum Chem 61:589
85. Nalewajski RF (2004) Chem Phys Lett 386:265
86. Gopinathan MS, Jug K (1983) Theor Chim Acta (Berl) 63:497
87. Nalewajski RF, Formosinho SJ, Varandas AJC, Mrozek J (1994) Int J Quantum Chem 52:1153
88. Mrozek J, Nalewajski RF, Michalak A (1998) Polish J Chem 72:1779
89. Jug K, Gopinathan MS (1990) In: Maksić ZB (ed) Theoretical models of chemical bonding, vol II. Springer, Heidelberg, p 77
90. Mayer I (1983) Chem Phys Lett 97:270
91. Nalewajski RF (2012) Int J. Quantum Chem (in press)
92. Esquivel RO, Flores-Gallegos N, Iuga C, Carrera E, Angulo JC, Antolín J (2009) Theor Chem Acc 124:445
93. López-Rosa S (2010) Information-theoretic measures of atomic and molecular systems, PhD Thesis, University of Granada
94. López-Rosa S, Esquivel RO, Angulo JC, Antolín J, Dehesa JS, Flores-Gallegos N (2010) J Chem Theory Comput 6:145
95. Luken WL, Beratan DN (1984) Theoret Chim Acta (Berl) 61:265
96. Luken WL, Culberson JC (1984) Theoret Chim Acta (Berl) 66:279
97. Putz MV (2005) Int J Quantum Chem 105:1
98. Shaik S, Danovich D, Wu W, Hiberty PC (2009) Nat Chem 1:443

Struct Bond (2012) 149: 95–130
DOI: 10.1007/978-3-642-32753-7_3

Applying Conventional Ab Initio and Density Functional Theory Approaches to Electric Property Calculations. Quantitative Aspects and Perspectives

George Maroulis

Abstract We have examined the predictive capability of density functional theory methods in calculations of electric polarizability and hyperpolarizability. We have focused on test cases belonging to three high-priority classes of molecular systems: "soft" metal clusters, novel types of compounds, and weakly bonded molecules. The performance of theoretical methods over arbitrary collections of molecular properties can be analyzed and classified by the introduction of a new methodology based on graph theoretic considerations and pattern recognition techniques.

Keywords Ab initio · Density functional theory · Electric (hyper)polarizability

Contents

G. Maroulis (✉)
Department of Chemistry, University of Patras, Patras 26500, Greece
e-mail: maroulis@upatras.gr

1 Introduction

The theory of electric polarizability is of fundamental importance to the rational approach and interpretation of large classes of phenomena [1]. In particular, these properties are of fundamental importance to intermolecular interaction studies [2], nonlinear optics [3], collision-induced spectroscopy [4], and the simulation of fluids [5, 6]. They are also routinely associated with general molecular characteristics as hardness [7], softness [8], hypersoftness [9], stiffness [10], and compressibility [11]. Understandably, polarizability is also linked to reactivity [12]. Another important field is QSAR, QSPR studies, and the understanding of pharmacological activity [13, 14].

In view of the important applications, the theoretical determination of electric properties of atoms, molecules, clusters, and even larger molecular architectures is rapidly expanding. The predictive capability of theoretical methods and convergence to the available experimental data has been closely examined in comprehensive reviews [15].

Two wide classes of theoretical methods are preferentially applied to the determination of electric polarizabilities: ab initio methods and density functional theory (DFT)-based approaches. Ab initio methods have been known to be converging reliably, displaying a high level of agreement to experimentally determined quantities. Very accurate ab initio treatments of electric polarizabilities are available for atoms and relatively small molecules. The distinct advantage of DFT methods lies in the possibility of economical, in a computational sense, treatments of relatively large molecular architectures. There, their advantage ends. It is usually very hard to determine the predictive capability of DFT-based methods or just to reasonably compare their performance to the presumably more accurate ab initio methodologies.

In this paper, we investigate the possibilities offered by widely used DFT methods. We have chosen test cases in three different, difficult classes of problems: (1) the linear and nonlinear polarizabilities of metal clusters, (2) the polarizabilities of novel compounds, and (3) the interaction-induced polarizability in weakly bonded systems. In particular, the three test cases are the sodium tetramer, a particularly "soft" molecule, the new compound HXeI, and the interaction polarizability of two water molecules in the dimer $(H_2O)_2$.

In previous work we employed an information theoretic approach to classify and systematically improve theoretical descriptions of molecules, introduced as arbitrary collections of data/properties [16, 17]. This approach relies on the availability of reference or accurate theoretical descriptions and is now easily accessible [18]. More recently, we have developed a more subtle, general approach based on graph theoretic arguments and pattern recognition techniques [19]. We rely on generalized metrics to introduce distance/proximity, order, and classification in spaces of theoretical descriptions. In addition, we introduce clustering in such spaces by the construction of a unique mathematical object, the minimum spanning tree (MST), and the performance of single-linkage cluster analysis (SLCA).

2 Theoretical Considerations

2.1 Electric Properties of Atoms, Molecules, and Clusters. Basic Theory and Computational Aspects

Our use of the theory of electric polarizability follows in all aspects, including the basic theoretical philosophy and terminology, the classic papers of Buckingham [20] and McLean and Yoshimine [21].

The energy (E^{p}) and perturbed electric moments ($\mu_\alpha{}^{\mathrm{p}}$, $\Theta_{\alpha\beta}{}^{\mathrm{p}}$, $\Omega_{\alpha\beta\gamma}{}^{\mathrm{p}}$) of an uncharged molecule in a weak, static electric field can be expanded as

$$\begin{aligned} E^{\mathrm{p}} &\equiv E^{\mathrm{p}}(F_\alpha, F_{\alpha\beta}, F_{\alpha\beta\gamma}, F_{\alpha\beta\gamma\delta}, \ldots) \\ &= \mathbf{E}^0 - \boldsymbol{\mu}_\alpha F_\alpha - (1/3)\boldsymbol{\Theta}_{\alpha\beta} F_{\alpha\beta} - (1/15)\boldsymbol{\Omega}_{\alpha\beta\gamma} F_{\alpha\beta\gamma} - (1/105)\boldsymbol{\Phi}_{\alpha\beta\gamma\delta} F_{\alpha\beta\gamma\delta} + \ldots \\ &\quad - (1/2)\boldsymbol{\alpha}_{\alpha\beta} F_\alpha F_\beta - (1/3)\mathbf{A}_{\alpha,\beta\gamma} F_\alpha F_{\beta\gamma} - (1/6)\mathbf{C}_{\alpha\beta,\gamma\delta} F_{\alpha\beta} F_{\gamma\delta} \\ &\quad - (1/15)\mathbf{E}_{\alpha,\beta\gamma\delta} F_\alpha F_{\beta\gamma\delta} + \ldots \\ &\quad - (1/6)\boldsymbol{\beta}_{\alpha\beta\gamma} F_\alpha F_\beta F_\gamma - (1/6)\mathbf{B}_{\alpha\beta,\gamma\delta} F_\alpha F_\beta F_{\gamma\delta} + \ldots \\ &\quad - (1/24)\boldsymbol{\gamma}_{\alpha\beta\gamma\delta} F_\alpha F_\beta F_\gamma F_\delta + \ldots, \end{aligned} \tag{1}$$

$$\begin{aligned} \mu_\alpha^p &= \boldsymbol{\mu}_\alpha + \boldsymbol{\alpha}_{\alpha\beta} F_\beta + (1/3)\mathbf{A}_{\alpha,\beta\gamma} F_{\beta\gamma} + (1/2)\boldsymbol{\beta}_{\alpha\beta\gamma} F_\beta F_\gamma + (1/3)\mathbf{B}_{\alpha\beta,\gamma\delta} F_\beta F_{\gamma\delta} \\ &\quad + (1/6)\boldsymbol{\gamma}_{\alpha\beta\gamma\delta} F_\beta F_\gamma F_\delta + \ldots, \end{aligned} \tag{2}$$

$$\Theta_{\alpha\beta}^p = \Theta_{\alpha\beta} + \mathbf{A}_{\gamma,\alpha\beta} E_\gamma + \mathbf{C}_{\alpha\beta,\gamma\delta} F_{\gamma\delta} + (1/2)\mathbf{B}_{\gamma\delta,\alpha\beta} F_\gamma F_\delta + \ldots, \tag{3}$$

$$\Omega_{\alpha\beta\gamma}^p = \boldsymbol{\Omega}_{\alpha\beta\gamma} + \mathbf{E}_{\delta,\alpha\beta\gamma} F_\delta + \ldots, \tag{4}$$

where the variables F_α, $F_{\alpha\beta}$, $F_{\alpha\beta\gamma}$, etc., are the field, field gradient, etc. at the origin of the molecule. The terms **in bold** are the permanent properties of the system: energy ($\mathbf{E}^0$), dipole ($\boldsymbol{\mu}_\alpha$), quadrupole ($\boldsymbol{\Theta}_{\alpha\beta}$), octopole ($\boldsymbol{\Omega}_{\alpha\beta\gamma}$), and hexadecapole ($\boldsymbol{\Phi}_{\alpha\beta\gamma\delta}$) moment. The second-, third-, and fourth-order properties are the dipole polarizability ($\boldsymbol{\alpha}_{\alpha\beta}$), dipole–quadrupole polarizability ($\mathbf{A}_{\alpha,\beta\gamma}$), quadrupole polarizability ($\mathbf{C}_{\alpha\beta,\gamma\delta}$), dipole–octopole polarizability ($\mathbf{E}_{\alpha,\beta\gamma\delta}$), first dipole hyperpolarizability ($\boldsymbol{\beta}_{\alpha\beta\gamma}$), dipole–dipole–quadrupole hyperpolarizability ($\mathbf{B}_{\alpha\beta,\gamma\delta}$), and second dipole hyperpolarizability ($\boldsymbol{\gamma}_{\alpha\beta\gamma\delta}$). The subscripts denote Cartesian components and a repeated subscript implies summation over x, y, and z. The number of independent components needed to specify the above tensors is strictly regulated by symmetry. In addition to the Cartesian components, of interest are the various invariants of some tensors. For the dipole polarizability and second dipole hyperpolarizability the mean is defined as

$$\bar{\alpha} = (1/3)\alpha_{\alpha\alpha} \text{ and } \bar{\gamma} = (1/5)\gamma_{\alpha\alpha\beta\beta}. \tag{5}$$

For some systems of simple symmetry, as linear molecules, the invariants of dipole (hyper)polarizability tensors, mean values, and anisotropies are easily defined and commonly used. Most of them are measurable quantities or can be deduced from experimental observations.

$$\begin{aligned}
\bar{\alpha} &= (\alpha_{zz} + 2\alpha_{xx})/3 \\
\Delta\alpha &= \alpha_{zz} - \alpha_{xx} \\
\bar{\beta} &= (3/5)(\beta_{zzz} + 2\beta_{zxx}) \\
\Delta\beta &= \beta_{zzz} - 3\beta_{zxx} \\
\bar{\gamma} &= (3\gamma_{zzzz} + 8\gamma_{xxxx} + 12\gamma_{xxzz})/15 \\
\Delta_1\gamma &= 3\gamma_{zzzz} - 4\gamma_{xxxx} + 3\gamma_{xxzz} \\
\Delta_2\gamma &= \gamma_{zzzz} + \gamma_{xxxx} - 6\gamma_{xxzz}.
\end{aligned} \tag{6}$$

When sufficiently weak electric fields are applied it is possible to extract the electric properties of the molecule from the above expansions. In previous work, we have applied various computational schemes based on the finite-field [22] approach to the calculation of electric properties from perturbed atomic/molecular energies and induced multipole moments [23–28].

Various computational aspects of the theoretical determination of electric properties are available in books [29, 30] or comprehensive collections [31–33].

All calculations reported in the following sections have been performed with quantum chemical methods easily accessible via the widely used GAUSSIAN suite of programs. See GAUSSIAN 98 [34] and GAUSSIAN 03 [35]. This arsenal includes conventional ab initio methods and density functional theory (DFT) approaches. We will not give further details about the structure and predictive capability of these methods here. For the interested reader, extensive presentations of the above classes of methods are clearly presented in standard references [36–39].

The ab initio methods used are:

- *SCF*, self-consistent-field
- *MP2, MP3, and MP4*: second-, third-, and fourth-order Møller–Plesset perturbation theory
- *DQ-MP4* and *SDQ-MP4*, partial fourth-order Møller–Plesset
- *CCSD*, singles and doubles coupled cluster
- *CCSD(T)*, which includes an estimate of connected triples via a perturbational treatment. This is the method with the, presumably, highest predictive potential

The DFT methods include the widely used B3LYP, B3PW91, and mPW1PW91 and many more that have been occasionally employed in electric property calculations.

Basis sets in modern quantum chemistry is too broad a subject to be examined in detail here [40]. The search for Gaussian basis sets suitable for molecular property calculations is vital to computational quantum chemistry [41]. Many computational schools have reported significant work in the direction of basis set construction [42–44] and testing of their performance. The construction of purpose-oriented basis sets has attracted particular attention [45–49]. Early work by Dykstra and coworkers [50, 51] or Spackman [52] has elucidated many important aspects of the performance of basis sets in electric property calculations. In our computational efforts we favor the use of molecule-specific, purpose-oriented basis sets. The construction of molecule-specific, purpose-oriented basis sets for large molecular architectures or low-symmetry polyatomics is largely impractical. Nevertheless, previous work shows that for systems of reasonable size as atoms [23, 53], atomic anions or cations [54, 55], diatomics [56, 57] and triatomics [58–61], symmetric polyatomics [62–64], and clusters [65–68], one can easily control the construction of the basis set in order to obtain suitable basis sets for electric property calculations.

2.2 Interaction-Induced Electric Properties

Interaction electric properties, as dipole moment, polarizability, and hyperpolarizability are of fundamental importance to the analysis and interpretation of measurements and observations in collision- and interaction-induced spectroscopies [69, 70]. Considerable progress has been recorded in recent years either on theoretical issues or computational advances. We single out a few significant papers related to the above field. Głaz et al. [71] reported a study of the collision-induced hyper-Rayleigh light-scattering spectra of He–Ne atomic pairs. Chrysos et al. [72] reported a study of the CO_2–Ar collision-induced ν_3 CO_2 band. A study of fundamental importance for collision-induced spectroscopy of gaseous CO_2, the determination of the exact low-order classical moments in collision-induced bands of linear rotors, was reported by Chrysos et al. [73]. A new treatment of the collision-induced Raman scattering by Ne–Ne was reported by Chrysos et al. [74]. Baranowska et al. [75] reported a theoretical study of the interaction-induced dipole moment and polarizability of CO–Ne. El-Kader et al. [76] determined the contributions of multipolar polarizabilities to the isotropic and anisotropic light scattering induced by intermolecular interactions in gaseous CH_4. Zvereva-Loëte et al. [77] and Buldakov et al. [78] reported an extensive study on the dipole moment surface and dipole polarizability surface for the CH_4–N_2 complex, a system of importance for the atmospheric physics of Titan. The calculation of the interaction-induced dipole moment, polarizability, and first and second hyperpolarizability of the H_2O–Rg (Rg $=$ He, Ne, Ar, Kr and Xe) complexes was reported by Haskopoulos and Maroulis [79]. Hartmann et al. [80] reported an investigation of the far-infrared collision-induced absorption band in gaseous CO_2.

Our work on the interaction-induced polarizability and hyperpolarizability of the water dimer relies on the conventional supermolecule approach. Details on this approach have been given in previous work on interaction-induced electric properties. In this approach, the interaction-induced properties of the $A \cdots B$ supermolecule are obtained as

$$P_{int}(A \cdots B) = P(A \cdots B) - P(A) - P(B). \tag{7}$$

In practice, the above equation is nearly exact when very large, nearly saturated basis sets are used. When truncated basis sets are used the basis set superposition error is removed by the counterpoise-correction method of Boys and Bernardi [81]. Thus, the above equation is replaced by

$$P_{int}(A \cdots B) = P(A \cdots B) - P(A \cdots X) - P(X \cdots B), \tag{8}$$

where $P(A \cdots X)$ denote calculation of the property for subsystem A in the presence of the ghost orbitals of subsystem B and $P(X \cdots B)$ for subsystem B in the presence of the ghost orbitals of subsystem A. It is easily shown that as the flexibility of the basis set increases one approaches a nearly ideal situation where $P(A \cdots X) \approx P(A)$ and $P(X \cdots B) \approx P(B)$.

For all computational aspects of the interaction-induced (hyper)polarizability of the water dimer, we lean heavily on previous experience of systems as CO_2–Rg [82], Ne–Ar [83], Xe–Xe [84], Kr–Xe [85], He–Ar [86], H_2–Ar [87], H_2–Ne [88], Kr–He [89], Kr–Ne [90], and H_2–Ne [91].

2.3 Proximity, Similarity, and Order in Spaces of Theoretical Descriptions

The evaluation of the performance of theoretical methods in atomic/molecular property calculations has been recognized as a formidable problem early enough. The quantification of the relative merit of theoretical methods is an essential part of modern computational quantum chemistry. Sometime ago we presented a methodology that relies on graph theoretic arguments and pattern recognition techniques to introduce order, classification, and clustering in spaces of arbitrary theoretical descriptions of atomic/molecular systems [92]. Our theory uses metric considerations to define distance/proximity and similarity in such spaces. The utility of such a methodology has been brought forth in various applications. Pattern recognition has long found application in Chemistry [93]. Our use of such techniques extends the application of pattern recognition (PR) to a hitherto unreached field: computational quantum chemistry (CQC). The logical strength of the analogy is made obvious by the following diagram:

Pattern recognition (PR)		Computational quantum chemistry (CQC)
Object	↔	Method
Features	↔	Molecular property values
Pattern	↔	Theoretical description (TD)
Pattern space	↔	Space of all TD

Thus in PR one has *objects* and in CQC *methods*. Objects are characterized by *features* and methods by *molecular property values*. A collection of features is a *pattern*; a collection of molecular property values a *theoretical description* (TD). A collection of patterns forms the pattern space, a collection of theoretical descriptions the *space of all theoretical descriptions.*

The above-mentioned methodology has found application in various quantum chemistry problems [94–96]. We give here only a few essential points and definitions.

Let TD_i be a collection of properties $Q_{\mathrm{m}\alpha}$ where the index m denotes methods and the index α denotes properties. The two indices take values in the index sets I_α and I_{m} (respectively).

$$\mathrm{TD}_i = \{Q_{\mathrm{m}\alpha}, m \in I_{\mathrm{m}}, \alpha \in I_\alpha\}. \tag{9}$$

We denote by TD the space of all theoretical descriptions TD_i:

$$\mathrm{TD} = \{\mathrm{TD}_1, \mathrm{TD}_2, \ldots \mathrm{TD}_N\} \quad \text{where} \quad i = 1, 2, \ldots N. \tag{10}$$

We define a generalized distance in the space of all TD by using the Minkowski metric. The distance between two theoretical descriptions TD_i and TD_j is defined as

$$D_{ij} \equiv D(\mathrm{TD}_i, \mathrm{TD}_j) = \left(\sum_\alpha \frac{|Q_{i\alpha} - Q_{j\alpha}|^p}{\left(\max\limits_{ij} |Q_{i\alpha} - Q_{j\alpha}| \right)^p} \right)^{1/p}, p \geq 1, 1 \leq i, j \leq N. \tag{11}$$

In most applications we have used the Euclidean metric, that is $p = 2$.

Similarity between two theoretical descriptions TD_i and TD_j is then defined on the basis of distance/proximity as

$$S_{ij} = 1 - \frac{D_{ij}}{\max\limits_{ij} D_{ij}}, \quad 1 \leq i, j \leq N. \tag{12}$$

By definition, $0 \leq S_{ij} \leq 1$.

To make a connection with graph theory we need a few standard definitions and interpretations. A reliable source of basic graph theory is the work of Chartrand and Lesniak [97]. A graph G is a finite nonempty set of objects called vertices together with a set of unordered pairs of vertices called edges. The vertex set of G is denoted **V(G)** and the edge set **E(G)**. The cardinality **p** of $V(G)$ is called the order of G and the cardinality **q** of $E(G)$ the size of G. Let us consider the Cartesian product TD × TD. The *graph of theoretical descriptions* G_{TD} has as vertex set $V(G_{\mathrm{TD}})$ the set of theoretical descriptions TD_i. The edge set $E(G_{\mathrm{TD}})$ is a subset of above defined Cartesian product, $E(G_{\mathrm{TD}}) \subset \mathrm{TD} \times \mathrm{TD}$.

We assign to each edge of G_{TD} a real number, a *weight*. Consider the edge defined by TD_i and TD_j. We assign to the edge $\{TD_i, TD_j\}$ the real number $D_{ij} \equiv D(\mathrm{TD}_i, \mathrm{TD}_j)$. G_{TD} is now a *weighted graph.*

The diameter Diam G_{TD} of the graph of TD is defined as

$$\text{Diam} \quad G_{\text{TD}} = \max_{i,j \in V(G_{\text{TD}})} D_{ij} \equiv \max_{\text{TD}_i, \text{TD}_j \in V(G_{\text{TD}})} D\left(\text{TD}_i, \text{TD}_j\right). \tag{13}$$

In view of the above, the definition of the similarity can now be written as

$$S_{ij} \equiv S\left(\text{TD}_i, \text{TD}_j\right) = 1 - \frac{D\left(\text{TD}_i, \ \text{TD}_j\right)}{\text{Diam} \quad G_{\text{TD}}}, \text{TD}_i, \text{TD}_j \in V(G_{\text{TD}}). \tag{14}$$

Two more definitions are particular useful at this stage. The distance of fixed vertex u from a subset of the vertex set $S \subseteq V(G_{\text{TD}})$ is defined as

$$d(u, S) = \min_{x \in S}\{d(u, x)\}. \tag{15}$$

A point of major importance to our methodology is the definition of a *minimum spanning tree* (MST). A *spanning subgraph H* of a graph G has vertex and edge sets $V(H) \subseteq V(G)$ and $E(H) \subseteq E(G)$ and is of the same order as G. A graph G of order p and size q is a tree if and only if it is acyclic and $p = q + 1$. The weight of a spanning tree in a connected graph is the sum of the weights of its edges. Thus a *minimum spanning tree* of G is a *spanning tree* of G of minimum weight.

Suitable algorithms exist for the construction of a MST [98].

Here is now a synoptic form of our methodology step by step:

1. Define the space of theoretical descriptions TD.
2. Calculate all distances between vertices or TD_i of the graph G_{TD}.
3. Calculate the distance and similarity matrix.
4. Construct a weighted minimum spanning tree *MST* using known algorithms.
5. Perform a *single linkage cluster analysis* (SLCA) by removing all edges from the *MST* characterized by weights above a given threshold D_T.

This analysis creates a partition of the *MST* in distinct clusters $C_1, C_2, \ldots, C_K$. The union of all clusters is the vertex set of the G_{TD} graph: $V(G_{\text{TD}}) = \bigcup_{1 \le i \le K} C_i$.

Last, it is important to introduce a distance between clusters for our analysis. The nearest neighbor distance between clusters C_m and C_n is defined as

$$D_{nn}(C_m, C_n) = \min_{\text{TD}_i \in C_m, \text{TD}_j \in C_n} D(\text{TD}_i, \text{TD}_j). \tag{16}$$

3 Results

3.1 Sodium Tetramer, a Very Soft Molecule

The electric dipole polarizability of sodium clusters has been extensively studied, both experimentally [99–101] and theoretically [102–104]. In addition, the polarizability of the sodium atom is accurately known. The latest experimental value is

α(Na) = 162.7 ± 0.8, reported by Ekstroem et al. [105]. Reliable theoretical values have been reported both for the Hartree-Fock limit and higher levels of theory. The most accurate theoretical value α(Na) = 162.88 ± 0.6 has been reported by Thakkar and Lupinetti [106]. Consequently, reference values are readily available for the differential polarizability (DP) and the differential-per-atom-polarizability (DPA), defined as [66, 107]:

$$\mathrm{DP} \equiv \bar{\alpha}_{\mathrm{diff}} = (\bar{\alpha}(\mathrm{Na}_n) - n\alpha(\mathrm{Na})),$$

$$\mathrm{DPA} \equiv \bar{\alpha}_{\mathrm{diff}}/n = (\bar{\alpha}(\mathrm{Na}_n) - n\alpha(\mathrm{Na}))/n = (\bar{\alpha}(\mathrm{Na}_n))/n - \alpha(\mathrm{Na}).$$

The respective hyperpolarizability quantities, differential hyperpolarizability (DH) and the differential-per-atom-hyperpolarizability (DHPA), are defined as

$$\mathrm{DH} \equiv \bar{\gamma}_{\mathrm{diff}} = (\bar{\gamma}(\mathrm{Na}_n) - n\gamma(\mathrm{Na})),$$

$$\mathrm{DHPA} \equiv \bar{\gamma}_{\mathrm{diff}}/n = (\bar{\gamma}(\mathrm{Na}_n) - n\gamma(\mathrm{Na}))/n = (\bar{\gamma}(\mathrm{Na}_n))/n - \gamma(\mathrm{Na}).$$

DH and DHPA are even more basis set and method sensitive than DP and DPA. As it has been shown elsewhere the above quantities are very useful as rigorous criteria for the analysis of the performance of quantum chemical methods [108, 109].

We have chosen as testing ground for our theoretical treatment the sodium tetramer (Na_4). Most of the data pertaining to Na_4 and used in this section have been published elsewhere [110]. The molecular geometry of the tetramer is a rhombus with side 3.64 Å and a short diagonal of 3.30 Å. The molecule is placed on the *xz* plane with the four Na atoms at (0,0,±1.65) and (±3.244549,0,0). All electric property calculations have been performed with a purpose-oriented molecule-specific basis set of [7s5p2d] type. The theoretical methods used are conventional ab initio and a selection of DFT approaches. The ab initio methods are SCF, MP2, MP3, DQ-MP4, SDQ-MP4, MP4, CCSD, and CCSD(T). The choice of DFT methods leans heavily on previous experience [111–115]: B3LYP, B3PW91, PBEPBE, PBEPW91, mPW1PW91, and mPW1PBE as implemented in the GAUSSIAN suite of codes.

The independent Cartesian components of the polarizability ($\alpha_{\alpha\beta}$) and hyperpolarizability ($\gamma_{\alpha\beta\gamma\delta}$) are given in Table 1. The method-dependence of the mean and the anisotropy of the dipole polarizability and the mean second hyperpolarizability are shown in Figs. 1, 2, and 3, respectively. Figure 1 shows clearly that electron correlation reduces the size of the mean dipole polarizability. The sequence $\bar{\alpha}$(SCF) < $\bar{\alpha}$(MP2) < $\bar{\alpha}$(MP3) < $\bar{\alpha}$(MP4) shows a monotonic decrease of the values calculated with Møller–Plesset perturbation theory. Our best theoretical value $\bar{\alpha}$(CCSD(T)) is quite close to $\bar{\alpha}$(MP2) and $\bar{\alpha}$(MP3). For comparison, we show on the same figure the DFT values for the mean calculated with mPW1PW91 and B3PW91. The former is close to the MP4 value and the latter close to SDQ-MP4. Figure 2 shows that the method-dependence of the dipole polarizability

Table 1 Static (hyper)polarizability of Na_4 calculated with ab initio and DFT methods ($10^{-3} \times \gamma_{\alpha\beta\gamma\delta}$). All quantities in atomic units

Method	α_{xx}	α_{yy}	α_{zz}	γ_{xxxx}	γ_{yyyy}	γ_{zzzz}	γ_{xxyy}	γ_{yyzz}	γ_{xxzz}
SCF	894.87	356.17	449.18	16448	2563	3236	3369	881	3465
MP2	875.48	338.02	435.61	12941	2377	3153	2740	840	2721
MP3	870.31	337.28	434.84	11030	2222	2976	2368	796	2271
DQ-MP4	861.83	336.70	433.86	10443	2090	2789	2176	753	2037
SDQ-MP4	856.81	336.39	433.53	10036	1950	2597	2022	704	1861
MP4	862.06	336.75	435.39	10340	2016	2703	2109	728	1950
CCSD	855.83	336.83	436.57	11337	2035	2732	2247	729	2275
CCSD(T)	869.53	337.39	439.50	11700	2082	2833	2357	749	2414
B3LYP	835.22	299.40	395.43	12250	1990	2655	2722	733	2579
B3PW91	881.89	323.89	419.28	9946	1889	2614	2457	689	2466
PBEPBE	859.05	311.76	406.94	11143	1883	2640	2642	712	2511
PBEPW91	854.05	309.38	404.83	11309	1874	2614	2634	709	2500
mPW1PW91	884.63	327.59	421.52	9143	1868	2524	2366	694	2340
mPW1PBE	889.76	329.96	423.63	8754	1878	2548	2361	699	2334

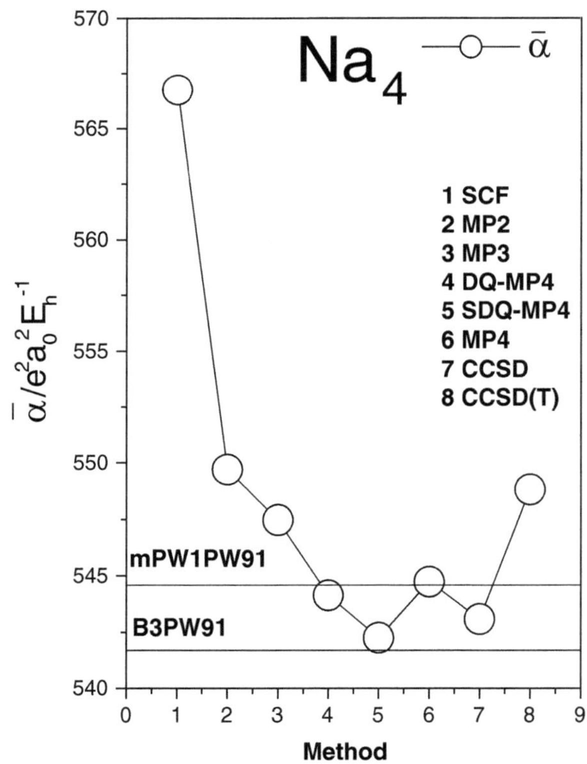

Fig. 1 Method dependence of the mean dipole polarizability of Na_4

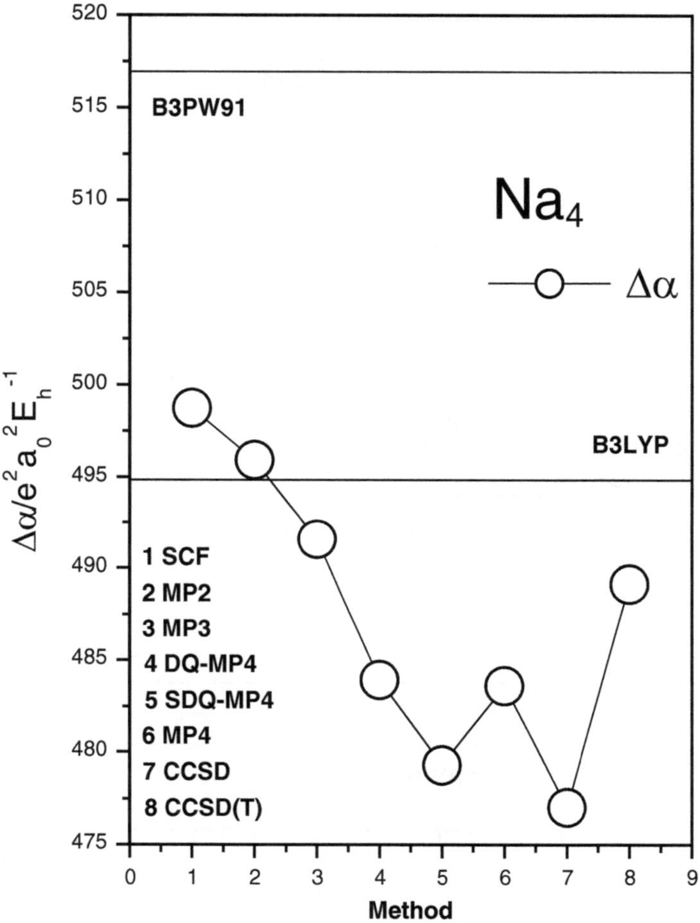

Fig. 2 Method dependence of the anisotropy the dipole polarizability of Na_4

anisotropy values bear close resemblance to that of the mean. We observe again the monotonic $\Delta\alpha$(SCF) < $\Delta\alpha$ (MP2) < $\Delta\alpha$ (MP3) < $\Delta\alpha$ (MP4). Agreement with the DFT methods is less obvious in this case. $\Delta\alpha$(B3LYP) is quite close to $\Delta\alpha$(MP2) but $\Delta\alpha$(B3PW91) is significantly higher than the ab initio values. Inspecting Fig. 3, we are surprised to find that the various orders of Møller–Plesset perturbation theory display once more a monotonic behavior for the mean second dipole hyperpolarizability: $\bar{\gamma}$(SCF) < $\bar{\gamma}$(MP2) < $\bar{\gamma}$(MP3) < $\bar{\gamma}$(MP4). $\bar{\gamma}$(MP3) is quite close to our, presumably, most accurate $\bar{\gamma}$ (CCSD) and $\bar{\gamma}$ (CCSD(T)). What is more three conventional DFT methods as B3LYP, B3PW91 and mPW1PW91 yield mean second hyperpolarizability values close enough to the high-level ab initio ones.

We have used the data in Table 1 to calculate distance/proximity and similarities for the ab initio and DFT theoretical descriptions (TD) of the electric (hyper) polarizability of the sodium tetramer. We have a total of nine independent Cartesian components so we have in our hands a 9D problem. The similarities between the

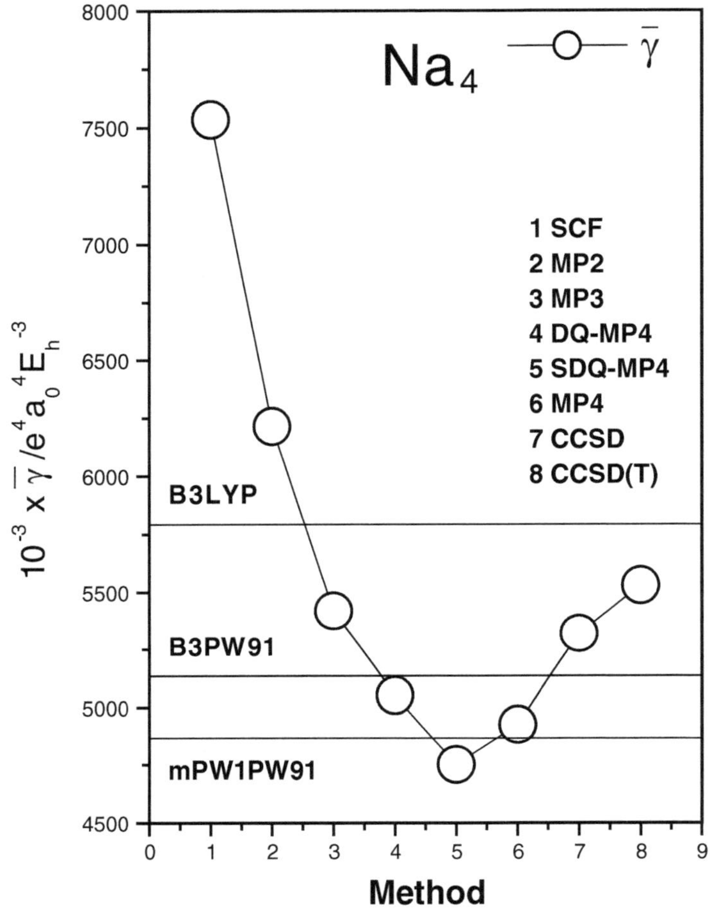

Fig. 3 Method dependence of the mean second hyperpolarizability of Na_4

theoretical descriptions (TD) of Table 1 are given in Table 2. The quantities shown in Table 2 allow us a direct quantitative evaluation of the similarity of the performance of two arbitrary methods over the calculation of the (hyper)polarizability of the sodium tetramer. In Fig. 4 we show a histogram with the similarity of any method to the presumably most accurate one CCSD(T). As this is our reference value, S(CCSD(T), CCSD(T)) = 1. The method closest to CCSD(T) is CCSD: S(CCSD, CCSD(T)) = 0.86337. Among the DFT methods, the closest to the reference is B3PW91, with S(B3PW91,CCSD(T)) = 0.68609. We are also able to glean more specific information about the relative performance of the various methods. Of all DFT methods, B3PW91 is closest to mPW1PW91, S(B3PW91, mPW1PW91) = 0.90949. PBEPBE is closest to PBEPW91, S(PBEPBE, PBEPW91) = 0.95325 and mPW1PW91 closest to mPW1PBE, S(mPW1PBE, mPW1PW91) = 0.94854.

The minimum spanning tree (**MST**) of the theoretical descriptions of Table 1 is shown in Fig. 5. The tree is a Graph $G(p,q)$ of order 14 and magnitude 13

Table 2 Similarity of the performance of theoretical methods on the electric (hyper)polarizability of Na_4

	SCF	MP2	MP3	DQ-MP4	SDQ-MP4	MP4	CCSD	CCSD(T)	B3LYP	B3PW91	PBEPBE	PBEPW91	mPW1PW91	mPW1PBE
SCF	1	0.57488	0.33884	0.16830	0.00280	0.09796	0.16868	0.28289	0	0.07459	0.05613	0.02549	0.02893	0.04121
MP2	0.57488	1	0.74056	0.55296	0.36835	0.46939	0.52730	0.63584	0.32852	0.40102	0.3940	0.36280	0.35548	0.36370
MP3	0.33884	0.74056	1	0.80334	0.61266	0.71629	0.74301	0.83139	0.42900	0.58075	0.53971	0.50707	0.55275	0.56012
DQ-MP4	0.16830	0.55296	0.80334	1	0.80770	0.90759	0.88283	0.84868	0.48311	0.67442	0.61870	0.59192	0.66610	0.66250
SDQ-MP4	0.00280	0.36835	0.61266	0.80770	1	0.89422	0.81645	0.70972	0.46637	0.68950	0.61724	0.60095	0.70972	0.69245
MP4	0.09796	0.46939	0.71629	0.90759	0.89422	1	0.88070	0.80267	0.47075	0.70139	0.62366	0.59946	0.70706	0.69923
CCSD	0.16868	0.52730	0.74301	0.88283	0.81645	0.88070	1	0.86337	0.51992	0.69631	0.65619	0.63300	0.67626	0.66199
CCSD(T)	0.28289	0.63584	0.83139	0.84868	0.70972	0.80267	0.86337	1	0.46914	0.68609	0.61677	0.58336	0.65589	0.65614
B3LYP	0	0.32852	0.42900	0.48311	0.46637	0.47075	0.51992	0.46914	1	0.54306	0.76642	0.80058	0.48531	0.44434
B3PW91	0.07459	0.40102	0.58075	0.67442	0.68950	0.70139	0.69631	0.68609	0.54306	1	0.77031	0.73162	0.90949	0.88024
PBEPBE	0.05613	0.39400	0.53971	0.61870	0.61724	0.62366	0.65619	0.61677	0.76642	0.77031	1	0.95325	0.70809	0.66765
PBEPW91	0.02549	0.36280	0.50707	0.59192	0.60095	0.59946	0.63300	0.58336	0.80058	0.73162	0.95325	1	0.67413	0.63113
mPW1PW91	0.02893	0.35548	0.55275	0.66610	0.70972	0.70706	0.67626	0.65589	0.48531	0.90949	0.70809	0.67413	1	0.94854
mPW1PBE	0.04121	0.36370	0.56012	0.66250	0.69245	0.69923	0.66199	0.65614	0.44434	0.88024	0.66765	0.63113	0.94854	1

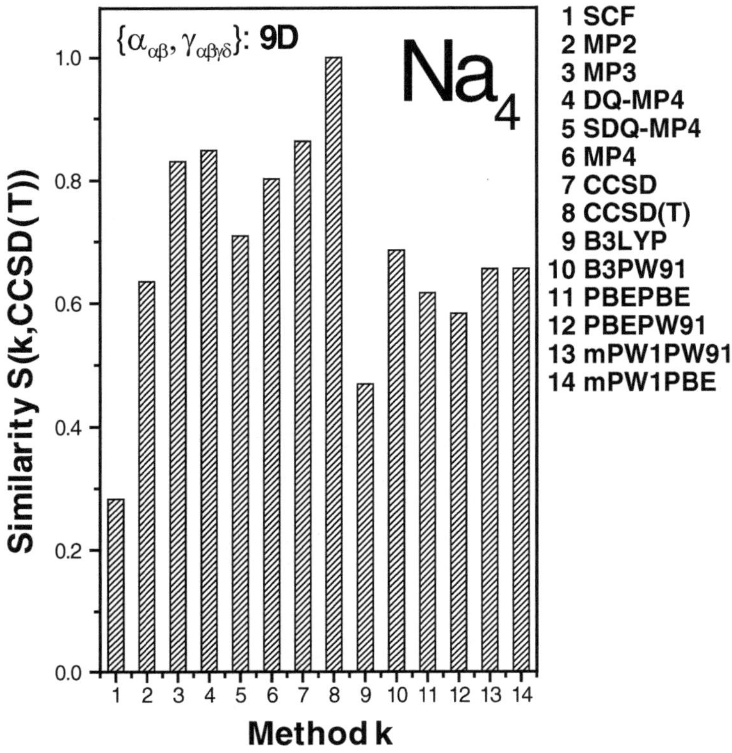

Fig. 4 Histogram of the similarities S(k,CCSD(T)) where k = ab initio and DFT for the 9D descriptions $\{\alpha_{xx}, \alpha_{yy}, \alpha_{zz}, \gamma_{xxxx}, \gamma_{yyyy}, \gamma_{zzzz}, \gamma_{xxyy}, \gamma_{yyzz}, \gamma_{zzxx}\}$

($p = q + 1$). Removing all edges $D(i,j)$ above a threshold of $D_T = 0.4$, $D(i,j) > D_T$ results in the clustering shown also in Fig. 5. The obtained clusters are as follows:

C1 = {SCF}
C2 = {MP2}
C3 = {MP3}
C4 = {DQ-MP4, SDQ-MP4, MP4, CCSD, CCSD(T)}
C5 = {B3LYP}
C6 = {PBEPBE, PBEPW91}
C7 = {B3PW91, mPW1PW91, mPW1PBE}

$$\mathrm{TD} = \mathrm{C1} \cup \mathrm{C2} \cup \mathrm{C3} \cup \mathrm{C4} \cup \mathrm{C5} \cup \mathrm{C6} \cup \mathrm{C7}$$

High-level ab initio results form a large cluster C4 = {DQ-MP4, SDQ-MP4, MP4, CCSD, CCSD(T)} which contains the most accurate theoretical descriptions CCSD and CCSD(T). The DFT methods form three distinct clusters C5, C6, and

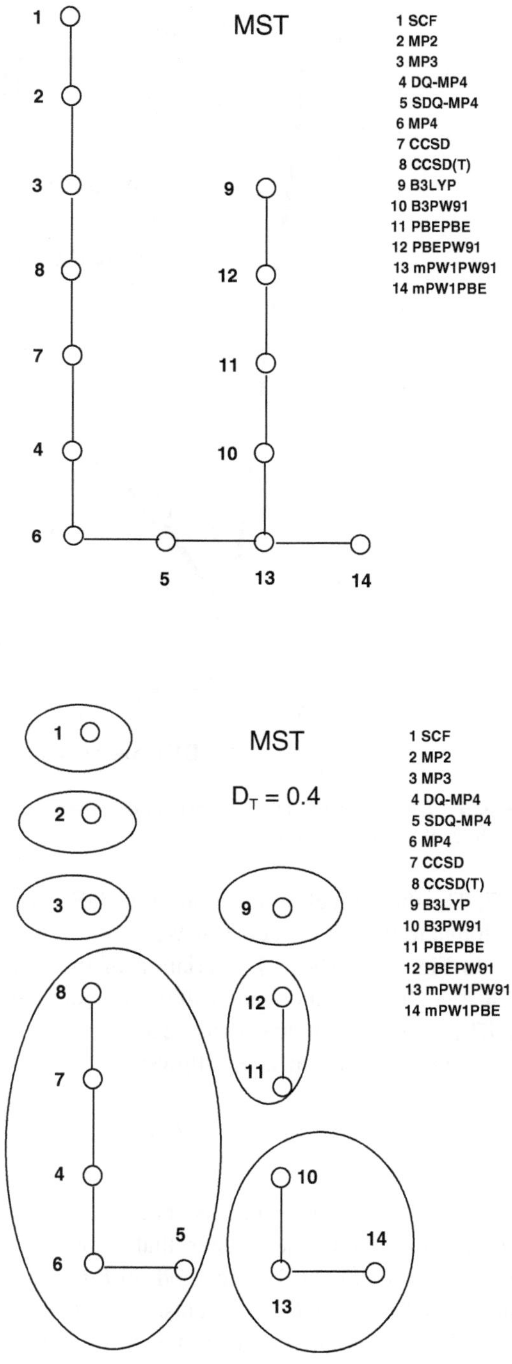

Fig. 5 Minimum spanning tree (*MST*) for the theoretical descriptions of the sodium tetramer and subsequent clustering with a distance threshold $D_{\mathrm{T}} = 0.4$.

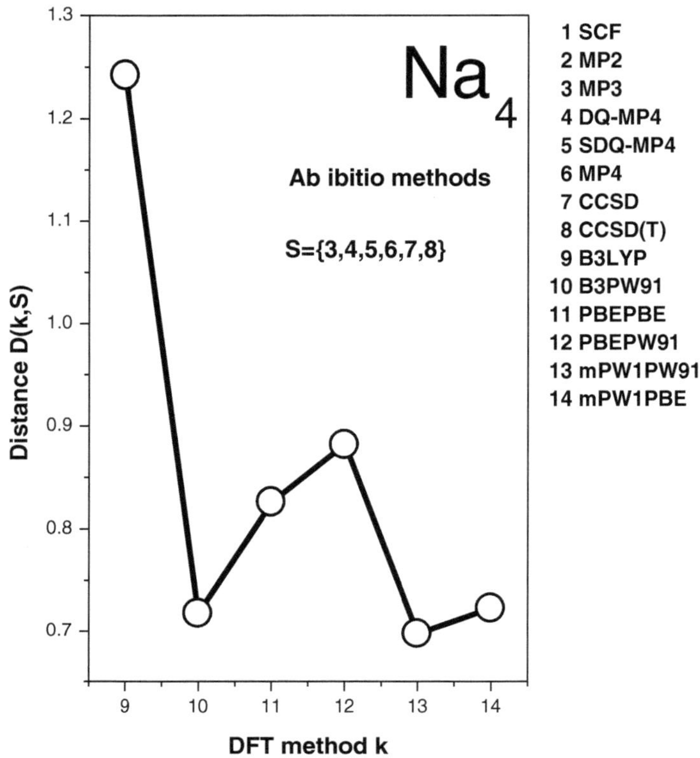

Fig. 6 Distance of the DFT theoretical descriptions from the C4 ≡ S cluster that contains the high-level ab initio methods

C7. The theoretical description of all DFT closest to CCSD(T) is B3PW91 which belongs to the C7 cluster, B3PW91 ∈ C7.

In Fig. 6 we show the variation of the distance of all DFT methods from the C4 ≡ *S* cluster that contains all high-level ab initio methods. Instead of comparing DFT methods individually to one ab initio theoretical description we define directly an element-set distance as follows:

$$D(x,S) = \min_{s \in S}(x,s)\,.$$

Here, $D(x,S)$ defines the distance of a theoretical description x from the S cluster. In this case S is the cluster that contains all high-level ab initio theoretical descriptions and x a DFT method. In Fig. 6 we see clearly the distance of all DFT methods from the C4 ≡ S cluster. This is a very realistic view of the highly complex matter of the performance of DFT methods. Figure 6 reveals that the DFT method most distant from the S cluster is the B3LYP one. The respective distance is D(B3LYP,S) = 1.242. The methods most close to the S cluster are

B3PW91 and mPW1PW91. The respective distances are D(B3PW91,S) = 0.7175 and D(mPW1PW91,S) = 0.6975. Ordering the DFT methods with respect to their distance from S results in the sequence D(B3LYP,S) > D(PBEPW91,S) > D(PBEPBE,S) > D(mPW1PBE,S) > D(B3PW91,S) > D(mPW1PW91,S).

3.2 New Classes of Molecules, the Case of HXeI

HXeI is a typical representative of a fascinating new class of molecules. HRgX compounds are produced by UV radiation of hydrogen halides (HX) in rare gas (RG) matrices [116]. They have attracted considerable experimental and theoretical attention [117]. The importance of the HXeI molecule in particular was brought forth by the work of Buck and Farnik [118]. In their experiments this linear molecule is detected by orientation in strong laser and weak electric fields [119]. Computational experience on HXeI is relatively limited. An empirical estimate of the anisotropy of the dipole polarizability has been proposed by Nahler et al. [120] in their work on the photodissociation of oriented HXeI molecules generated from HI–Xe_n clusters.

In this section we turn our attention on the electric dipole moment, polarizability, and hyperpolarizability of this important species. We lean heavily for molecular data and insights on our recent paper on the electric properties of HXeI [121].

We take into account two classes of molecular properties. Ab initio results calculated with basis sets (given in I/Xe/H order) B5 ≡ [10s9p8d1f/9s8p7d1f/6s3p1d] (179 CGTF) and B9 ≡ [11s10p10d3f/9s8p7d1f/7s5p1d] (214 CGTF). The B9 basis set was used in the DFT calculations. With basis B5 we calculated SCF, MP2, SDQ-MP4, MP4, CCSD, and CCSD(T) and with the larger B9 basis SCF and MP2 values. All ab initio values were taken from the above-cited paper. The DFT results were calculated with basis B9 and the methods B1LYP, B3LYP, B3PW91, mPW1PW91, HCTH, BHandH, BHandHLYP, PBEPBE, and PBEPW91. The molecular geometry of this linear molecule is defined by the bond lengths R(I–Xe) = 3.0577 and R(Xe–I) = 1.7077 Å. It was obtained at the MP2(Full)/B9 level of theory. At the same MP2/B9 level, a natural bond orbital analysis (NBO) yields charges of −0.55803 (I), 0.57691 (Xe), and −0.01888 (H). The calculated independent Cartesian components (z is the molecular axis) of the electric property tensors are given in Table 3.

The molecular values listed in Table 3 show clearly the difficulty in predicting reliable (hyper)polarizabilities for HXeI. This is particularly evident in the observed variations of the longitudinal components of the first and second hyperpolarizability. We base most of the presentation and analysis in this part on the invariants of (hyper)polarizability: mean values $\bar{\alpha}$, $\Delta\alpha$, $\bar{\beta}$, and $\bar{\gamma}$. Figure 7 shows the method dependence of the dipole moment. We also give a few characteristic DFT values. The high-level ab initio data are well grouped together. The DFT method closest to the most accurate CCSD(T)/B5 value is the B3PW91/B9 one which yields μ_z = 2.4181 ea_0. The BHandHLYP method predicts a dipole moment

Table 3 Electric properties of HXeI at the theoretical molecular geometry

Basis	Method	μ	α_{zz}	α_{xx}	β_{zzz}	β_{zxx}	γ_{zzzz}	γ_{xxxx}	γ_{xxzz}
B5	SCF	3.1642	158.1475	58.7304	−2788.7	179.5	159534	25857	7392
	MP2	2.6450	178.4193	61.3754	−2502.8	198.1	45910	33854	13524
	SDQ-MP4	2.6243	177.7069	61.1082	−2489.1	208.9	36556	32441	11775
	MP4	2.4836	181.7763	61.9752	−2111.3	205.5	−19509	35827	13888
	CCSD	2.6229	176.8633	60.9417	−2338.8	214.2	21232	31478	11800
	CCSD(T)	2.4584	181.2297	61.5717	−1858.2	220.2	−51560	33852	13652
B9	SCF	3.1621	158.1235	58.8893	−2776.3	182.4	157783	24797	6106
	MP2	2.6102	178.3887	61.5569	−2453.5	196.3	19795	8707	−2040
B9	B1LYP	2.4300	164.8947	60.9186	−1568.5	194.6	124931	36455	15993
	B3LYP	2.3857	164.4301	60.9369	−1487.3	193.0	124643	36478	16142
	B3PW91	2.4181	161.6319	59.8737	−1438.6	182.0	111561	32474	14202
	mPW1PW91	2.4670	161.3968	59.6325	−1492.7	179.9	108993	31914	13952
	HCTH	2.2235	163.5979	61.9986	−1344.2	206.4	164649	47502	20850
	BHandH	2.7623	162.1835	59.4488	−1970.6	175.6	111730	29621	11917
	BHandHLYP	2.6863	163.1892	59.2536	−1947.4	182.3	108529	29737	11884
	PBEPBE	2.2242	162.5679	61.9794	−1239.9	192.8	137383	40680	18205
	PBEPW91	2.2194	162.6710	62.0161	−1236.2	193.5	137304	40812	18287

All quantities in atomic units

close enough to the MP2 one. In Fig. 8 the evolution of the method dependence of the mean dipole polarizability shows a clear gap between SCF and the post-Hartree–Fock methods, a sign of a very strong electron correlation effect. We observe a clear discrepancy between DFT and ab initio methods. Among the most reliable DFT methods the HCTH and B3PW91 yield mean dipole polarizabilities $\bar{\alpha} = 95.87$ and $93.79\, e^2a_0^2E_h^{-1}$, respectively, both clearly below the most accurate CCSD and CCSD(T) results. The anisotropy of the dipole polarizability is normally a more severe test for DFT methods than that of the mean. In Fig. 9, the method dependence of the anisotropy resembles closely that of the mean. Among all other DFT methods, we note the performance of BHandHLYP, B3PW91, and PBE which give $\Delta\alpha = 103.94$, 101.76, and 100.59 $e^2{a_0}^2{E_h}^{-1}$, respectively, all three close enough to the SCF value. The ab initio values for the mean first hyperpolarizability are shown in Fig. 10. Of all DFT methods, BHandHLYP performs best, yielding a value $\bar{\beta} = -949.6\, e^3a_0^3E_h^{-2}$. This result is close to both MP4 and CCSD(T). The performance of other DFT methods is characteristically poor. B3PW91 yields $\bar{\beta} = -644.8\, e^3a_0^3E_h^{-2}$, a value significantly lower in magnitude than the BHandHLYP one. Last, ab initio and DFT values for the mean second hyperpolarizability are given in Fig. 11. Electron correlation lowers significantly the magnitude of this important property. With the notable exception of BH and HLYP and B3PW91, which predict values close to the SCF one, $\bar{\gamma} = -50993$ and $47073\, e^4a_0^4E_h^{-3}$, respectively, all other DFT methods yield considerably higher values.

We have calculated distance/proximity values for all methods/basis sets used in this section. The respective theoretical descriptions are 8D strings of molecular properties $\{\mu_z, \alpha_{zz}, \alpha_{xx}, \beta_{zzz}, \beta_{zxx}, \gamma_{zzzz}, \gamma_{xxxx}, \gamma_{xxzz}\}$. The calculated values are given

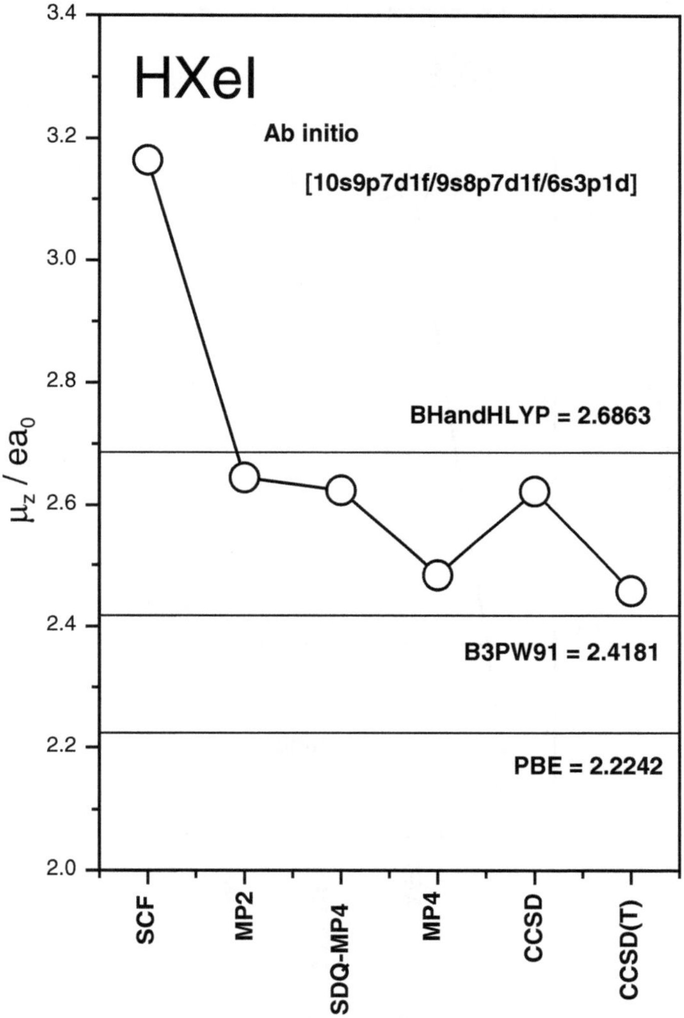

Fig. 7 Method dependence of the dipole moment of HXeI

in Table 4. To simplify matters, we have assigned numbers to all methods: 1 = SCF, 2 = MP2, 3 = SDQ-MP4, 4 = MP4, 5 = CCSD, 6 = CCSD(T), 7 = SCF, 8 = MP2, 9 = B1LYP, 10 = B3LYP, 11 = B3PW91, 12 = mPW1PW91, 13 = HCTH, 14 = BHandH, 15 = BHandHLYP, 16 = PBEPBE, and 17 = PBEPW91. To avoid confusion we sometimes denote methods 7 and 8 by 7 = SCF-B9 and 8 = MP2-B9, respectively, to avoid confusion with 1 = SCF and 2 = MP2 results that have been calculated with basis set B5. The most distant or dissimilar theoretical descriptions are 1 = SCF and 6 = CCSD(T): $S(1,6) = 0$. The two SCF/B5 and SCF/B9 descriptions are very similar, $S(1,7) = 0.95157$. This is not the case for the MP2/B5 and MP2/B9 descriptions as S

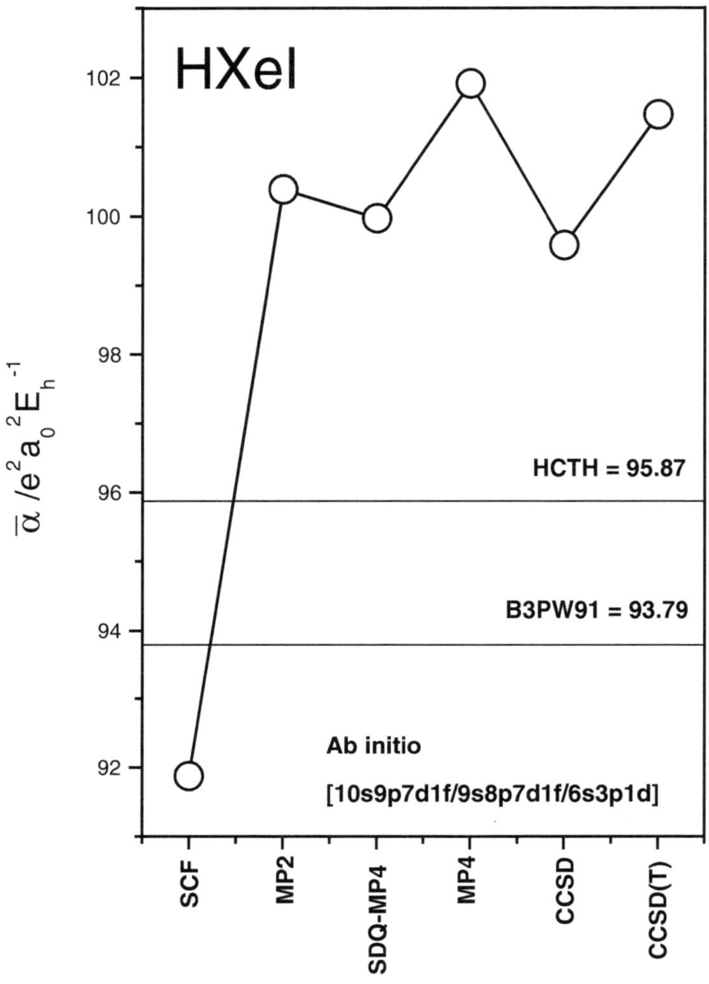

Fig. 8 Method dependence of the mean polarizability of HXeI

(2,8) = 0.55279. In Fig. 12 we show an histogram of the evolution of the similarity $S(k,6) \equiv S(k, \text{CCSD(T)})$. As 6 = CCSD(T) is the reference theoretical description here the maximum similarity for the histogram in Fig. 12 is $S(6,6) = 1$. The ab initio theoretical description closest to 6 = CCSD(T) is 4 = MP4: $S(4,6) = 0.80257$. From the DFT descriptions the closest to 6 = CCSD(T) is 9 = B1LYP: $S(6,9) = 0.41201$. Thus, as one easily gathers from the histogram, the similarity between ab initio (calculated with basis set B5) and DFT descriptions is rather poor. This is also the case for the comparison of ab initio results calculated with basis sets B5 and B9. To view similarity from another perspective, we show in Fig. 13 the evolution of the similarity $S(k,11) \equiv S(k, \text{B3PW91})$. The 11 = B3PW91 method is similar enough to MP2, SDQ-MP4, and CCSD. It is most close to the 12 = mPW1PW91 method: $S(11,12) = 0.94834$.

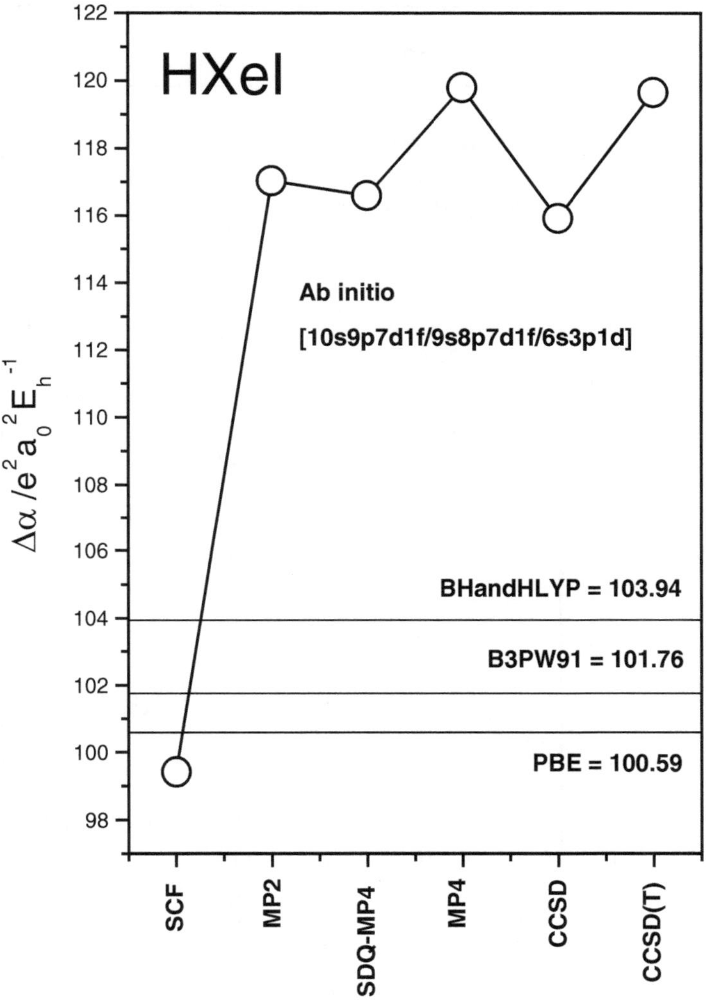

Fig. 9 Method dependence of the polarizability anisotropy of HXeI

The minimum spanning tree (MST) corresponding to the graph representing the theoretical descriptions (Table 3) is shown in Fig. 14. This tree is a graph $G(p,q)$ of order $p = 17$ and size $q = p - 1$. One expects this graph to be of sufficient complex structure. Clustering by removing all edges greater than a threshold value $D_T = 0.4$ results in the following partitioning of the spaces of theoretical descriptions:

$C1 \equiv \{1,7\}$
$C2 \equiv \{2,3,5\}$
$C3 \equiv \{4\}$
$C4 \equiv \{6\}$

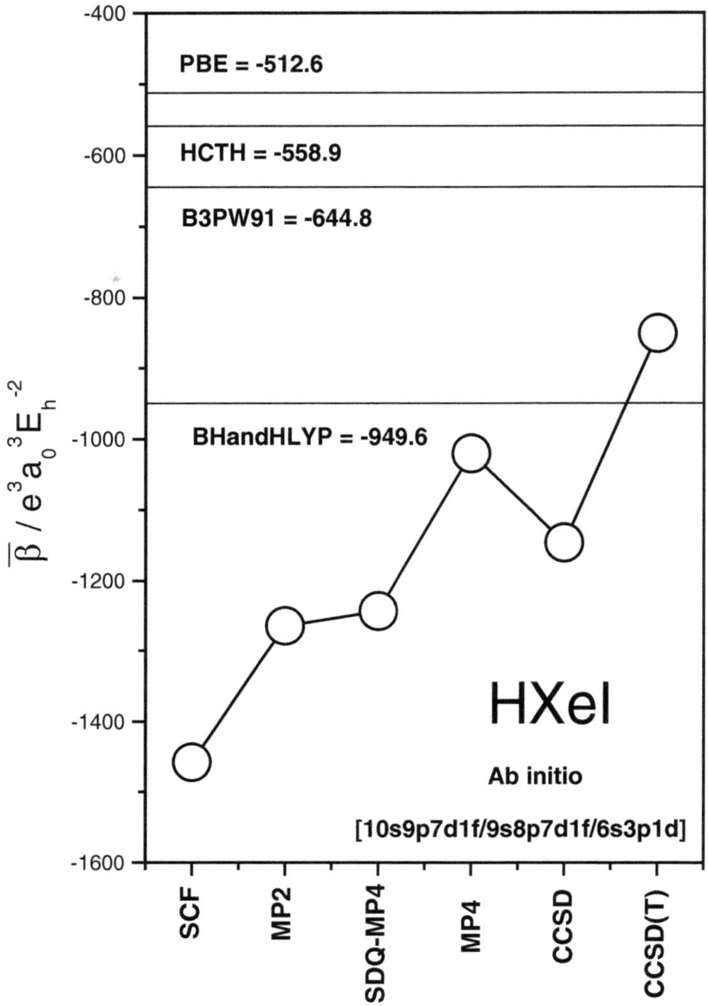

Fig. 10 Method dependence of the mean first hyperpolarizability of HXeI

C5 ≡ {8}
C6 ≡ {9,10}
C7 ≡ {11,12}
C8 ≡ {13,16,17}
C9 ≡ {14,15}

$$\mathrm{TD} = \mathrm{C1} \cup \mathrm{C2} \cup \mathrm{C3} \cup \mathrm{C4} \cup \mathrm{C5} \cup \mathrm{C6} \cup \mathrm{C7} \cup \mathrm{C8} \cup \mathrm{C9} \equiv \bigcup_i C_i, i = 1, 2, \ldots, 9$$

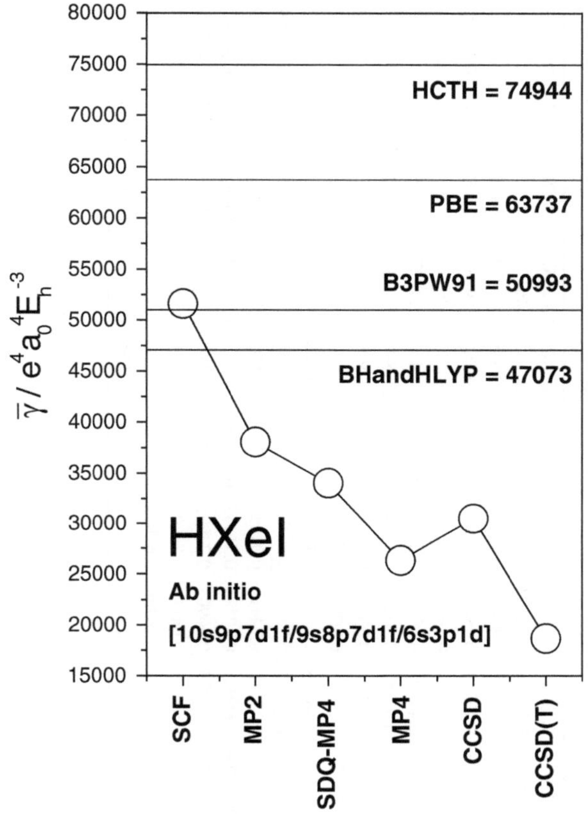

Fig. 11 Method dependence of the mean second hyperpolarizability of HXeI

Some characteristics of this clustering are easily discernible. The two SCF and SCF-B9 descriptions form cluster C1. Three high-level ab initio theoretical descriptions form cluster C2 ≡ {2,3,5} ≡ {MP2, SDQ-MP4, CCSD}. The composition of the C6, C7, and C8 clusters grouping DFT methods is clearly understood: B3PW91 is very close to mPW1PW91, BH and H is close to BH and HLYP, and PBEPBE is close to PBEPW91. HCTH belongs to the same cluster as PBEPBE and PBEPW91 and for good reason: it is most similar to these two DFT methods with S (HCTH,PBEPBE) ≈ S(HCTH,PBEPW91) ≈ 0.8.

Last, in Fig. 15 we show the evolution of the distance of the ab initio descriptions calculated with basis set B5 from the reference group of widely used B3LYP, B3PW91, and mPW1PW91. The distance $d(x,S)$, where S = {B3LYP, B3PW91, mPW1PW91} and $x \in \Omega \equiv$ {SCF, MP2, SDQ-MP4, MP4, CCSD, CCSD(T)}, is shortest for the most accurate CCSD(T) method:

$$d(CCSD(T), S) \leq \underset{x\in\Omega}{d(x,S)}\,.$$

Table 4 Similarity of the performance of theoretical methods on the electric (hyper)polarizability of HXeI

	1	2	3	4	5	6	7	8	9	10	11	12	13	14	15	16	17
1	1	0.28938	0.27216	0.07659	0.24691	0	0.95157	0.21474	0.31802	0.29578	0.38810	0.42457	0.06871	0.62357	0.59859	0.12434	0.11749
2	0.28938	1	0.87055	0.75645	0.79704	0.61276	0.30389	0.55279	0.54984	0.52172	0.43077	0.42155	0.37835	0.46129	0.48752	0.40343	0.40237
3	0.27216	0.87055	1	0.75570	0.91459	0.67538	0.29179	0.56777	0.52845	0.49821	0.40435	0.39383	0.36635	0.42352	0.46524	0.37277	0.37264
4	0.07659	0.75645	0.75570	1	0.75124	0.80257	0.09307	0.49525	0.47353	0.45494	0.33823	0.31800	0.36556	0.29868	0.33247	0.38519	0.38670
5	0.24691	0.79704	0.91459	0.75124	1	0.72627	0.26807	0.54984	0.52566	0.49619	0.40304	0.39116	0.36818	0.40579	0.45498	0.36943	0.36977
6	0	0.61276	0.67538	0.80257	0.72627	1	0.01971	0.42909	0.41201	0.39452	0.28160	0.25971	0.32456	0.21827	0.26705	0.32338	0.32609
7	0.95157	0.30389	0.29179	0.09307	0.26807	0.01971	1	0.24443	0.32761	0.30446	0.38960	0.42390	0.07897	0.61925	0.59852	0.13342	0.12682
8	0.21474	0.55279	0.56777	0.49525	0.54984	0.42909	0.24443	1	0.3193	0.30055	0.27217	0.26888	0.10919	0.32457	0.34357	0.18465	0.18269
9	0.31802	0.54984	0.52845	0.47353	0.52566	0.41201	0.32761	0.3193	1	0.96155	0.77561	0.73615	0.69901	0.62023	0.65465	0.77331	0.76842
10	0.29578	0.52172	0.49821	0.45494	0.49619	0.39452	0.30446	0.30055	0.96155	1	0.78799	0.74413	0.70584	0.60903	0.64114	0.79780	0.79211
11	0.38810	0.43077	0.40435	0.33823	0.40304	0.28160	0.38960	0.27217	0.77561	0.78799	1	0.94834	0.51457	0.74095	0.76795	0.62996	0.62168
12	0.42457	0.42155	0.39383	0.31800	0.39116	0.25971	0.42390	0.26888	0.73615	0.74413	0.94834	1	0.46906	0.78064	0.80443	0.58152	0.57313
13	0.06871	0.37835	0.36635	0.36556	0.36818	0.32456	0.07897	0.10919	0.69901	0.70584	0.51457	0.46906	1	0.33563	0.36696	0.81210	0.81885
14	0.62357	0.46129	0.42352	0.29868	0.40579	0.21827	0.61925	0.32457	0.62023	0.60903	0.74095	0.78064	0.33563	1	0.91223	0.43192	0.42385
15	0.59859	0.48752	0.46524	0.33247	0.45498	0.26705	0.59852	0.34357	0.65465	0.64114	0.76795	0.80443	0.36696	0.91223	1	0.45053	0.44312
16	0.12434	0.40343	0.37277	0.38519	0.36943	0.32338	0.13342	0.18465	0.77331	0.79780	0.62996	0.58152	0.81210	0.43192	0.45053	1	0.99007
17	0.11749	0.40237	0.37264	0.38670	0.36977	0.32609	0.12682	0.18269	0.76842	0.79211	0.62168	0.57313	0.81885	0.42385	0.44312	0.99007	1

1 = SCF
2 = MP2
3 = SDQ-MP4
4 = MP4
5 = CCSD
6 = CCSD(T)
7 = SCF-B9
8 = MP2-B9
9 = B1LYP
10 = B3LYP
11 = B3PW91
12 = mPW1PW91
13 = HCTH
14 = BHandH
15 = BHandHLYP
16 = PBEPBE
17 = PBEPW91

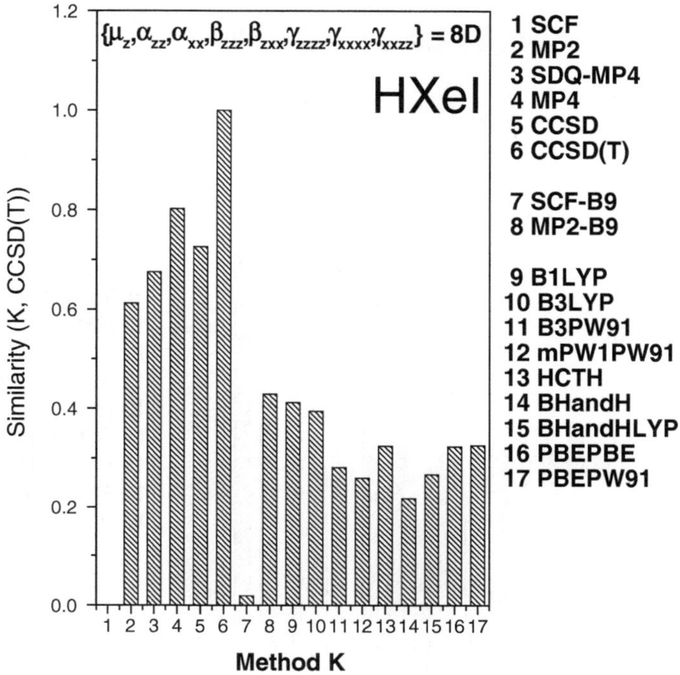

Fig. 12 Histogram of the similarities S(k,CCSD(T)) where k = ab initio and DFT for the 8D descriptions $\{\mu_z, \alpha_{zz}, \alpha_{xx}, \beta_{zzz}, \beta_{zxx}, \gamma_{zzzz}, \gamma_{xxxx}, \gamma_{xxzz}\}$

The distance $d(x,S)$ increases as follows:

$$d(\mathrm{CCSD}(T), S) < d(\mathrm{SCF}, S) < d(\mathrm{MP4}, S) < d(\mathrm{CCSD}, S) < d(\mathrm{SDQ-MP4}, S) < d(\mathrm{MP2}, S).$$

3.3 *Interaction-Induced Polarizability and Hyperpolarizability of Two Water Molecules*

In previous work [122] we presented an extended computational study of the interaction-induced electric properties of the water dimer $(H_2O)_2$. Our findings strongly suggest that the interaction-induced mean dipole polarizability and hyperpolarizability are nearly additive, as

$$\overline{\alpha}(H_2O)_2 \approx 2\overline{\alpha}(H_2O) \text{ and } \overline{\gamma}(H_2O)_2 \approx 2\overline{\gamma}(H_2O).$$

This surprising result appears to hold for the dipole polarizability of certain classes of water clusters, as brought forth in the work of Rodriguez et al. [123] or Ghanty and Ghosh [124]. The above defined additivity results clearly suggests that

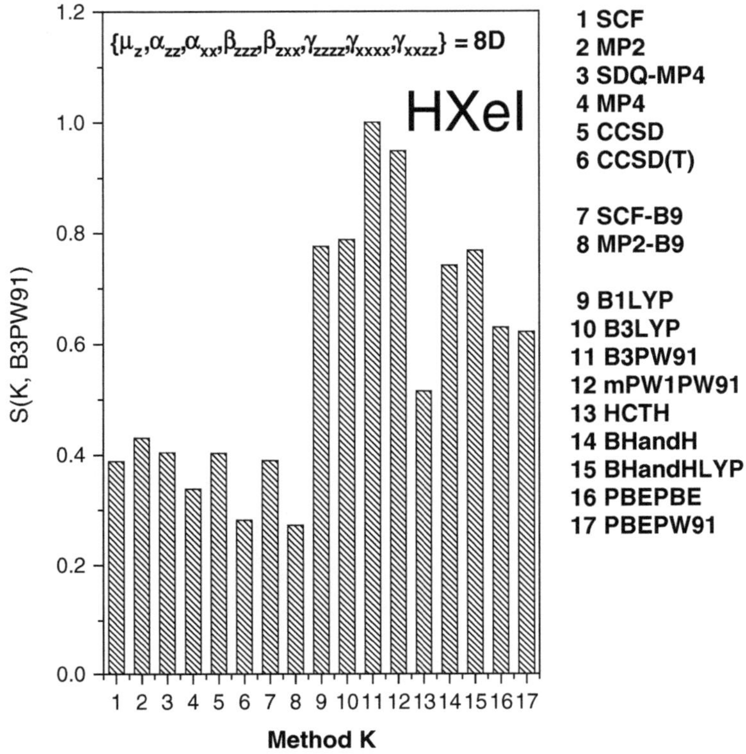

Fig. 13 Histogram of the similarities S(k,B3PW91) where k = *ab initio* and DFT for the 8D descriptions $\{\mu_z, \alpha_{zz}, \alpha_{xx}, \beta_{zzz}, \beta_{zxx}, \gamma_{zzzz}, \gamma_{xxxx}, \gamma_{xxzz}\}$

the respective interaction-induced mean (hyper)polarizability is very small. Identifying the proton acceptor (A) and proton donor (D) in the two moieties of $(H_2O)_2 \equiv AD$, we define the interaction properties in the counterpoise corrected (CP) scheme as

$$\overline{\alpha}_{\text{int}}(\text{H}_2\text{O})_2 = \overline{\alpha}(\text{H}_2\text{O})_2 - \overline{\alpha}(\text{H}_2\text{O}-\text{X}) - \overline{\alpha}(X-\text{H}_2\text{O}),$$
$$\overline{\gamma}_{\text{int}}(\text{H}_2\text{O})_2 = \overline{\gamma}(\text{H}_2\text{O})_2 - \overline{\gamma}(\text{H}_2\text{O}-\text{X}) - \overline{\gamma}(\text{X}-\text{H}_2\text{O}).$$

In a very recent paper, we demonstrated that the calculation of the interaction-induced (hyper)polarizability of a molecular system as the water dimer constitutes a sever test of the validity of DFT methods [125]. In this section we add more calculations in order to extend our observations to the performance of a larger class of DFT-based approximations.

Full computational details about the work presented in this section are given in our papers on the water dimer. See Fig. 16 for the relative orientation and respective role of the two water molecules. We adopt a dimer geometry that keeps the

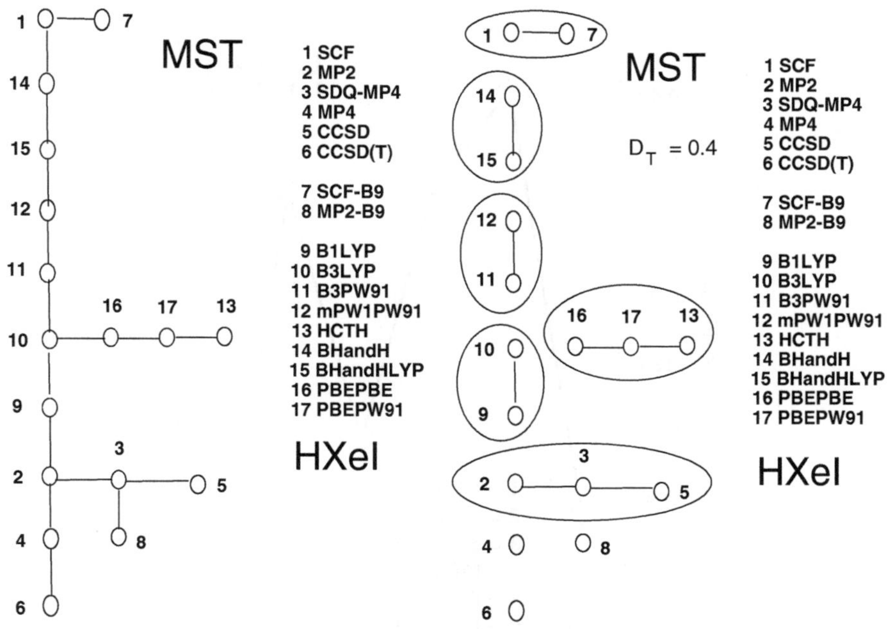

Fig. 14 Minimum spanning tree (MST) for the theoretical descriptions of HXeI and subsequent clustering with a distance threshold $D_T = 0.4$

monomer H_2O geometry frozen [126]. Thus, comparison of supermolecule to monomer properties makes eminent sense. All calculations presented here have been performed with the water-molecule-specific [6s4p3d1f/4s3p1d] basis set. The ab initio methods employed are SCF, MP2, SDQ-MP4, MP4, CCSD, and CCSD(T). The DFT methods are B1LYP, B3LYP, B3PW91, mPW1PW91, mPW1PBE, PBEPBE, PBEPW91, HCTH, BHandH, and BHandHLYP.

In Table 5 we give all monomer and dimer value used in the analysis of the results. The H_2O monomer (M) values pertain to the molecular geometry of the moiety in the dimer. The per-monomer (PM) values are simply defined as PM $=$ AD/2 that is $\overline{\alpha}(H_2O)_2/2$ for the polarizability and $\overline{\gamma}(H_2O)_2/2$ for the hyperpolarizability. The properties of interest are the interaction mean (hyper) polarizabilities obtained by the formula AD $-$ AX $-$ XD (see above). A first and valuable observation here is the strong disagreement of ab initio and DFT methods on the (hyper)polarizability of the water dimer. Such behavior for the DFT methods has been noted and analyzed early enough [127]. We must emphasize at this point the essential difference of the BHandH and BHandHLYP DFT methods. Both seem to be very close to the ab initio ones and quite distinct from the other DFT approaches.

In Fig. 17 we show the histogram of interaction-induced mean polarizability for all methods. With the notable exception of the BHandHLYP method the interaction property $\overline{\alpha}_{int}(H_2O)_2$ is negative for all ab initio methods and positive for the DFT

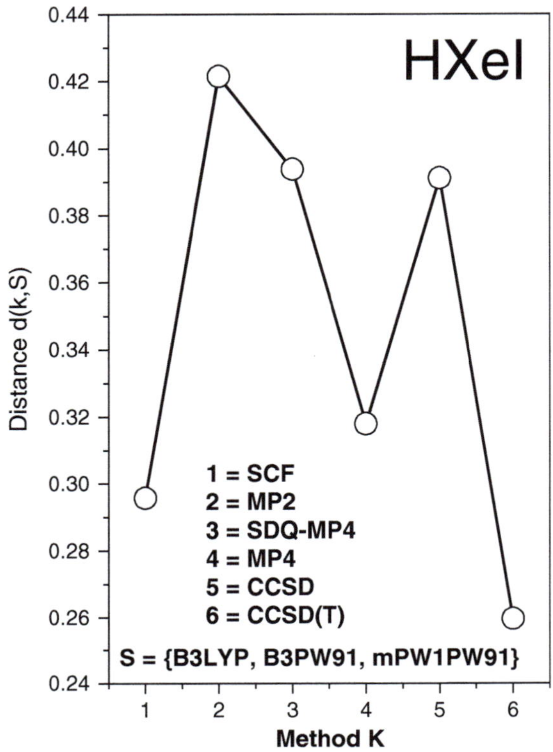

Fig. 15 Distance of the ab initio theoretical descriptions calculated with basis B5 from the reference DFT group S = {B3LYP, B3LYP, mPW1PW91}

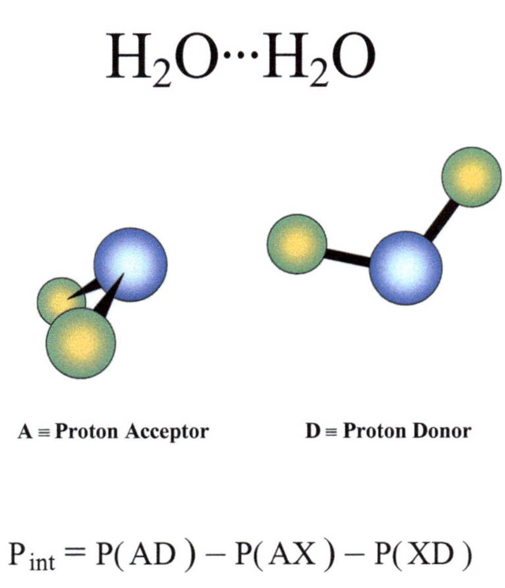

Fig. 16 Relative orientation of the two interacting water monomers in $(H_2O)_2$ and definition of the interaction-induced properties

Table 5 Analysis of the interaction-induced electric properties for the theoretical equilibrium molecular geometry of the water dimer (AD)

Property	Method	M	PM	AD	AX	XD	AD − AX − XD
$\overline{\alpha}$	SCF	8.48	8.44	16.89	8.49	8.49	**−0.09**
	MP2	9.85	9.81	19.62	9.84	9.85	**−0.07**
	SDQ-MP4	9.66	9.63	19.26	9.65	9.66	**−0.05**
	MP4	9.95	9.92	19.84	9.94	9.95	**−0.05**
	CCSD	9.58	9.56	19.13	9.58	9.58	**−0.03**
	CCSD(T)	**9.78**	**9.77**	**19.54**	**9.77**	**9.78**	**−0.01**
	B1LYP	9.85	9.87	19.74	9.86	9.86	**0.01**
	B3LYP	9.94	9.97	19.95	9.96	9.96	**0.03**
	B3PW91	9.78	9.81	19.62	9.79	9.79	**0.03**
	mPW1PW91	9.67	9.71	19.41	9.68	9.69	**0.04**
	mPW1PBE	9.67	9.71	19.41	9.69	9.69	**0.03**
	PBEPBE	10.54	10.62	21.25	10.56	10.56	**0.13**
	PBEPW91	10.53	10.62	21.24	10.55	10.55	**0.14**
	HCTH	10.28	10.35	20.71	10.30	10.30	**0.11**
	BHandH	9.27	9.30	18.60	9.28	9.28	**0.04**
	BHandHLYP	9.19	9.20	18.38	9.20	9.20	**−0.01**
$\overline{\gamma}$	SCF	975	957	1914	997	995	**−78**
	MP2	1742	1743	3486	1795	1790	**−99**
	SDQ-MP4	1698	1715	3429	1750	1746	**−67**
	MP4	1929	1952	3905	1992	1987	**−74**
	CCSD	1645	1671	3342	1690	1688	**−36**
	CCSD(T)	**1796**	**1834**	**3669**	**1847**	**1842**	**−20**
	B1LYP	2113	2309	4618	2249	2231	**137**
	B3LYP	2219	2449	4898	2366	2347	**185**
	B3PW91	2083	2260	4520	2215	2193	**113**
	mPW1PW91	1975	2145	4291	2102	2074	**115**
	mPW1PBE	1976	2116	4231	2079	2053	**99**
	PBEPBE	3090	3575	7150	3308	3266	**576**
	PBEPW91	3073	3557	7114	3288	3247	**578**
	HCTH	3115	3653	7306	3294	3307	**704**
	BHandH	1505	1571	3141	1569	1554	**19**
	BHandHLYP	1496	1529	3058	1537	1519	**1**

M ≡ Monomer, PM ≡ Per Molecule, AD ≡ $(H_2O)_2$
Reference values in *bold*. All quantities in atomic units

ones. An almost extreme behavior is obvious for the PBEPBE, PBEPW91, and HCTH methods.

The interaction-induced mean hyperpolarizabilities $\overline{\gamma}_{int}(H_2O)_2$ are shown in Fig. 18. The very same pattern as in the case of $\overline{\alpha}_{int}(H_2O)_2$ is visible here as well. The PBEPBE, PBEPW91, and HCTH methods yield very large values for the

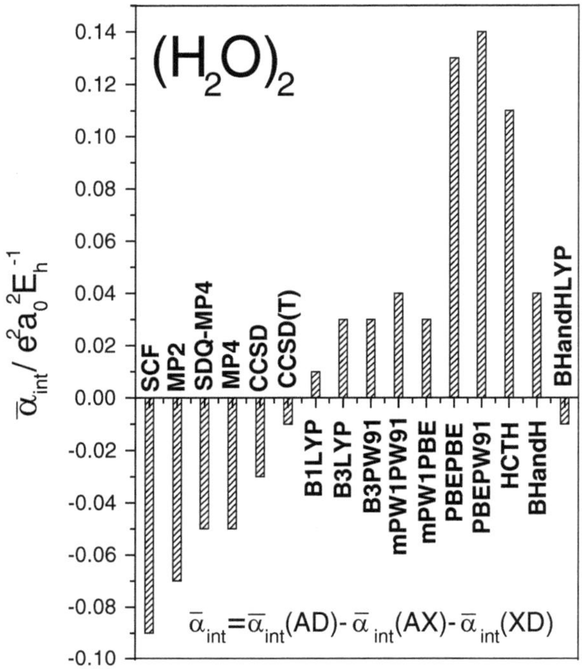

Fig. 17 Histogram of the performance of theoretical methods on the calculation of the interaction-induced mean dipole polarizability of the water dimer

interaction hyperpolarizability. The BHandH and BHandHLYP methods again give surprisingly small $\bar{\gamma}_{int}(H_2O)_2$ values.

Last, in Fig. 19 we show the evolution of the DFT values for the interaction-induced hyperpolarizability for the monomer (M) and the per-monomer (PM) value. In fact the $PM = \bar{\gamma}(H_2O)_2/2$ and $M = \bar{\gamma}(H_2O)$ values determine the differential-per-monomer hyperpolarizability, defined as

$$\mathrm{DHPM} = [\bar{\gamma}(H_2O)_2 - 2\bar{\gamma}(H_2O)]/2 = \bar{\gamma}(H_2O)_2/2 - \bar{\gamma}(H_2O) \equiv \mathrm{PM} - M.$$

In Fig. 19 we also show the reference M and P values calculated at the CCSD(T)/[6s4p3d1f/4s3p1d] level of theory. We observe that the sequence B3LYP, B3PW91, mPW1PW91, mPW1PBE displays a smooth convergence towards the reference CCSD(T) values. The PBEPBE, PBEPW91, and HCTH methods are characterized by very large M and PM values. Obviously, the BHandH and BHandHLYP methods appear as something of an anomaly here. Both the PM and M values for these two methods are lower than the respective reference CCSD(T) values.

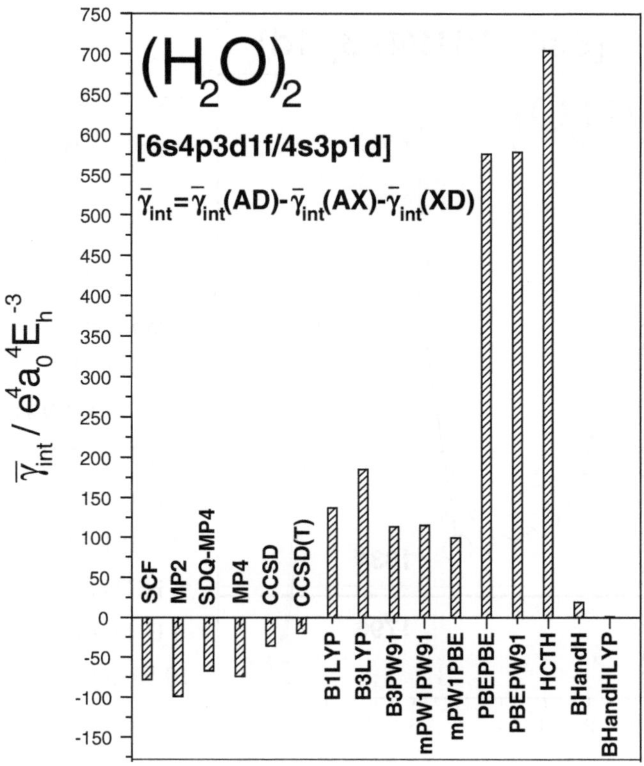

Fig. 18 Histogram of the performance of theoretical methods on the calculation of the interaction-induced mean second hyperpolarizability of the water dimer

4 Final Remarks and Conclusions

We have investigated and closely analyzed the predictive capability of DFT methods in the calculation of electric polarizabilities and hyperpolarizabilities for three difficult cases: the sodium tetramer, the new compound HXeI, and the interaction (hyper)polarizability of the two moieties in the weakly bonded water dimer. In the first two cases we have shown that the ab initio and DFT methods offer a quite different picture of the electric polarizability and hyperpolarizability. In the case of $(H_2O)_2$ we are led to two drastically divergent pictures as the two classes of methods clearly differ even in the sign of the interaction-induced mean polarizability and hyperpolarizability.

We show that it is probably a distinct advantage to talk about the quality or predictive capability of a set of DFT methods instead of trying to obtain a picture of the potential of a single one. This strongly corroborates our recent strategy to rely on a given set of DFT methods in order to form a clear idea about the DFT perspective in electric property calculations [115, 128, 129].

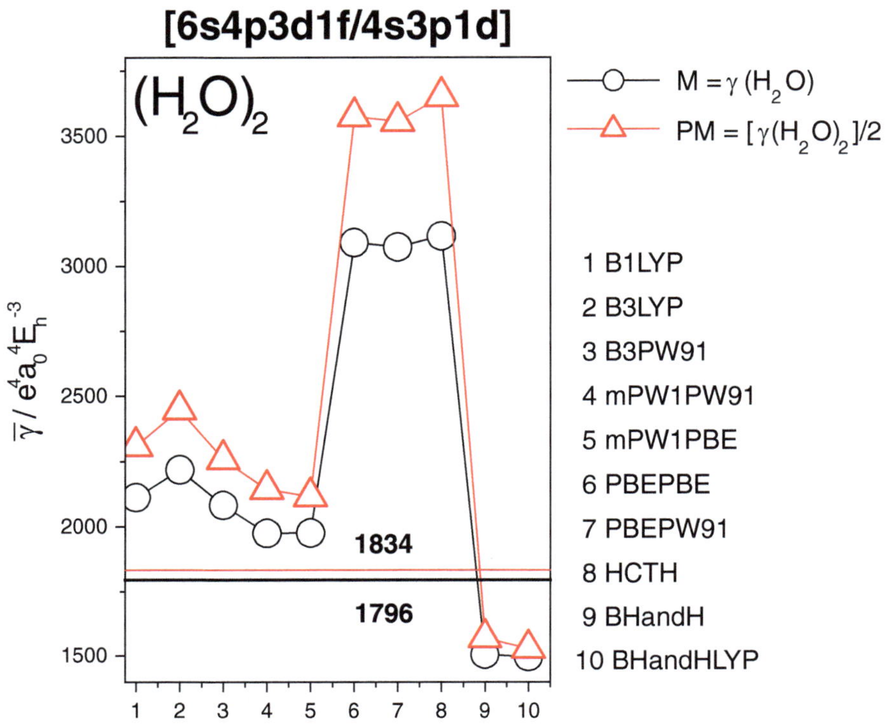

Fig. 19 Mean second hyperpolarizability for the water monomer (M) and dimer (per monomer PM) for all DFT methods. Reference CCSD(T) values are also included

References

1. Buckingham AD in Pullman B (ed) (1978) Intermolecular interactions: from diatomics to biopolymers, Wiley, New York, p. 1
2. Kaplan IG (2006) Intermolecular interactions. Wiley, Chichester
3. Hanna DC, Yuratich MA, Cotter D (1979) Nonlinear optics of free atoms and molecules. Springer, Berlin
4. Hartmann JM, Boulet C, Robert D (2008) Collisional effects on molecular spectra: laboratory experiments and models, consequences for applications. Elsevier, Amsterdam
5. Gray CG, Gubbins KE (1984) Theory of molecular fluids. Clarendon, Oxford
6. Gray CG, Gubbins KE, Joslin CG (2012) Theory of molecular fluids. volume 2: applications. Clarendon, Oxford
7. Berkowitz M, Parr RG (1988) J Chem Phys 88:2554
8. Vela A, Gázquez JL (1990) J Am Chem Soc 112:1490
9. Liu PH, Hunt KLC (1995) J Chem Phys 103:10597
10. Torrent-Succarat M, De Proft F, Geerlings P (2005) J Phys Chem A 109:6071
11. Donald KJ (2006) J Phys Chem A 110:2283
12. Matito E, Putz MV (2011) J Phys Chem A 115:12459–12462
13. Karelson K, Lobanov VS (1996) Phys Rev 96:1027–1043

14. Hansch C, Steinmetz WE, Leo AJ, Mekapati SB, Kurup A, Hoekman D (2003) J Chem Inf Comput Sci 43:120–125
15. Shelton DP, Rice JE (1994) Chem Rev 94:3–29
16. Maroulis G, Sana M, Leroy G (1981) Int J Quant Chem 19:43–60
17. Maroulis G (1988) Int J Quant Chem 24:185–190
18. Sordo JA (1988) Comp Phys Comm 113:85–104
19. Maroulis G, In: Sen KD (2002) Reviews of modern quantum chemistry. A celebration of Robert G. Parr, World Scientific, Singapore, pp 320–339
20. Buckingham AD (1967) Adv Chem Phys 12:107–142, and references therein
21. McLean AD, Yoshimine M (1967) J Chem Phys 47:1927–1935
22. Cohen HD, Roothaan CCJ (1965) J Chem Phys 43:S34–S39
23. Maroulis G, Bishop DM (1985) Chem Phys Lett 114:182–186
24. Bishop DM, Maroulis G (1985) J Chem Phys 82:2380–2391
25. Maroulis G, Thakkar AJ (1988) J Chem Phys 88:7623–7632
26. Maroulis G, Thakkar AJ (1988) J Chem Phys 89:7320–7323
27. Maroulis G (1991) J Chem Phys 94:1182–1190
28. Maroulis G (1998) J Chem Phys 108:5432–5448
29. Dykstra CE (1988) Ab initio calculation of the structures and properties of molecules. Elsevier, Amsterdam
30. Sauer SPA (2011) Molecular electromagnetism: a computational chemistry approach. Oxford University Press, Oxford
31. Maroulis G (ed) (2006) Atoms, molecules and clusters in electric fields: theoretical approaches to the calculation of electric polarizability. Imperial College Press, London
32. Maroulis G (ed) (2006) Computational aspects of electric polarizability calculations: atoms, molecules and clusters. Ios Press, Amsterdam
33. Papadopoulos MG, Sadlej AJ, Leszczynski J (eds) (2006) Non-linear optical properties of matter. from molecules to condensed phases. Springer, Berlin
34. Frisch MJ, Trucks GW, Schlegel HB et al (1998) GAUSSIAN 98, Revision A.7. Gaussian, Pittsburgh PA
35. Frisch MJ, Trucks GW, Schlegel HB et al (2004) GAUSSIAN 03, Revision D.01. Gaussian, Wallingford, CT
36. Szabo A, Ostlund NS (1982) Modern quantum chemistry. McMillan, New York
37. Wilson S (1984) Electron correlation in molecules. Clarendon, Oxford
38. Urban U, Cernusak I, Kellö V, Noga J (1987) Methods Comput Chem 1:117
39. Helgaker T, Jørgensen P, Olsen J (2000) Molecular electronic-structure theory. Wiley, Chichester
40. Wilson S (1987) Adv Chem Phys 67:439–500
41. Davidson ER, Feller D (1986) Chem Rev 86:681
42. Thakkar AJ, Koga T, Saito M, Hoffmeyer RE (1993) S27, 343–354
43. Koga T, Saito M, Hoffmeyer RE, Thakkar AJ (1994) JMolStruct (THEOCHEM) 306:249–260
44. Prascher BP, Woon DE, Peterson KA, Dunning TH Jr, Wilson AK (2011) Theor Chem Acc 128:69–82
45. Kellö V, Sadlej AJ (1995) Theor Chim Acta 91:353–371
46. Zuev MB, Nefediev SE (2004) J Comput Methods Sci Eng 4:481–491
47. Arruda PM, Canal Neto A, Jorge FE (2009) Int J Quant Chem 109:1189–1199
48. Baranowska AJ, Sadlej AJ (2010) J Comput Chem 31:552–560
49. de Berrêdo RC, Jorge FE, Jorge SS, Centoducatte R (2011) Comput Theor Chem 965:236–239
50. Liu SY, Dykstra CE (1987) J Phys Chem 91:1749–1754
51. Dykstra CE, Liu SY, Malik DJ (1989) Adv Chem Phys 75:37–111
52. Spackman MA (1989) J Phys Chem 93:7594–7603
53. Maroulis G, Bishop DM (1985) J Phys B 24:4675–4682

54. Maroulis G, Bishop DM (1986) Mol Phys 57:359–367
55. Maroulis G, Bishop DM (1986) J Phys B 19:369–377
56. Maroulis G, Bishop DM (1986) Mol Phys 58:273–283
57. Maroulis G, Makris C, Hohm U, Goebel D (1997) J Phys Chem A 101:953–956
58. Maroulis G (1994) J Chem Phys 101:4949–4955
59. Maroulis G (1992) Chem Phys Lett 199:250–256
60. Maroulis G, Pouchan C (1998) Phys Rev 57:2440–2447
61. Maroulis G (2003) Chem Phys 291:81–95
62. Maroulis G, Thakkar AJ (1991) J Chem Phys 95:9060–9064
63. Maroulis G (1996) Chem Phys Lett 259:654–660
64. Maroulis G (1996) J Chem Phys 105:8467–8468
65. Maroulis G, Xenides D (1999) J Phys Chem A 103:4590–4593
66. Maroulis G, Pouchan C (2003) Phys Chem Chem Phys 5:1992–1995
67. Karamanis P, Maroulis G, Pouchan C (2006) J Chem Phys 124:071101
68. Maroulis G, Karamanis P, Pouchan C (2007) J Chem Phys 126:154316
69. Birnbaum G (1985) Phenomena induced by intermolecular interactions. Plenum, New York
70. Tabisz GC, Neuman MN (eds) (1995) Collision- and interaction-induced spectroscopy. Kluwer, Dordrecht
71. Głaz W, Bancewicz T, Godet JL, Maroulis G, Haskopoulos A (2006) Phys Rev A 73:042708
72. Chrysos M, Rachet F, Egorova NI, Kouzov AP (2007) Phys Rev A 75:012707
73. Chrysos M, Kouzov AP, Egorova NI, Rachet F (2008) Phys Rev Lett 100:133007
74. Chrysos M, Dixneuf S, Rachet F (2009) Phys Rev A 80:054701
75. Baranowska A, Fernández B, Rizzo A, Jansik B (2009) Phys Chem Chem Phys 11:9871–9883
76. El-Kader, El-Sheikh, Bancewicz T, Hellmenn R (2009) J Chem Phys 131:044314
77. Zvereva-Loëte N, Kalugina YN, Boudon V, Buldakov MA, Cherepanov VN (2010) J Chem Phys 133:184302
78. Buldakov MA, Cherepanov VN, Kalugina YN, Zvereva-Loëte N, Boudon V (2010) J Chem Phys 132:164304
79. Haskopoulos A, Maroulis G (2010) J Phys Chem A 114:8730–8741
80. Hartmann JM, Boulet C, Jacquemart D (2011) J Chem Phys 134:094316
81. Boys Sf, Bernardi F (1970) 19:553–556
82. Maroulis G, Haskopoulos A (2001) Chem Phys Lett 349:335–341
83. Maroulis G, Haskopoulos A (2002) Chem Phys Lett 358:64–70
84. Maroulis G, Haskopoulos A, Xenides D (2004) Chem Phys Lett 396:59–65
85. Haskopoulos A, Xenides D, Maroulis G (2005) Chem Phys 309:271–275
86. Maroulis G, Haskopoulos A, Głaz W, Bancewicz T, Godet JL (2006) Chem Phys Lett 428:28–33
87. Bancewicz T, Głaz W, Godet JL, Maroulis G (2008) J Chem Phys 129:124306
88. Haskopoulos A, Maroulis G (2010) Chem Phys Lett 367:127–135
89. Xenides D, Hantzis A, Maroulis G (2011) Chem Phys 382:80–87
90. Chantzis A, Maroulis G (2011) Chem Phys Lett 507:42–47
91. Głaz W, Godet JL, Haskopoulos A, Bancewicz T, Maroulis G (2011) Phys Rev A 84:012503
92. Maroulis G (1995) Int J Quant Chem 55:173–180
93. Varmuza K (1980) Pattern recognition in chemistry. Springer, Heidelberg
94. Maroulis G (1999) J Chem Phys 111:583–591
95. Xenides D (2007) J Mol Struct (THEOCHEM) 804:41–46
96. Christodouleas C, Xenides D, Simos TE (2010) J Comput Chem 31:412–420
97. Chartrand G, Lesniak L (1986) Graphs and digraphs. Wadsworth, Belmont
98. Spath H (1980) Cluster analysis algorithms. Ellis Horwood, Chichester
99. Antoine R, Rayane D, Allouche AR, Aubert-Frecon M, Benichou E, Dalby FW, Dugourd PH, Broyer M, Guet C (1999) J Chem Phys 110:5568–5577
100. Tikhonov G, Kasperovich K, Wong K, Kresin VV (2001) Phys Rev A 64:063202

101. Bowlan J, Liang A, de Heer WA (2011) Phys Rev Lett 106:043401
102. Calaminici P, Jug K, Köster A (1999) J Chem Phys 111:4613–4620
103. Jiemchooroj A, Norman P, Semelius B (2006) J Chem Phys 125:124306
104. Aguado A, Vega A, Balbás LC (2011) Phys Rev B 84:165450
105. Ekstrom CR, Schmiedmayer J, Chapman MS, Hammond TD, Pritchard DE (1995) Phys Rev A 51:3883–3888
106. Thakkar AJ, Lupinetti C (2005) Chem Phys Lett 402:270–273
107. Maroulis G, Begué D, Pouchan C (2003) J Chem Phys 119:794–797
108. Maroulis G (2004) J Chem Phys 121:10519–10524
109. Papadopoulos MG, Reis H, Avramopoulos A, Erkoc S, Amirouche L (2005) J Phys Chem B 109:18822–18830
110. Xenides D, Maroulis G (2007) J Comput Methods Sci Eng 7:431–442
111. Karamanis P, Maroulis G (2005) Match Commun Math Comput Chem 53:269–282
112. Maroulis G, Haskopoulos (2009) J Comput Theor Nanosci 6:418–427
113. Maroulis G (2009) J Comput Theor Nanosci 6:886–893
114. Maroulis G (2011) Theor Chem Acc 129:437–445
115. Karamanis P, Maroulis G (2011) J Phys Org Chem 24:588–599
116. Petterson M, Lundell J, Räsänen M (1995) J Chem Phys 102:6423–6431
117. Petterson M, Lundell J, Räsänen M (1999) Eur J Inorg Chem 729–737 and references therein
118. Buck U, Farnik M (2006) Int Rev Phys Chem 25:583–612, and references therein
119. Buck U (2002) J Phys Chem A 106:10049–10062
120. Nahler NH, Baumfalk R, Buck U, Bihary Z, Benny Gerber R, Friedrich B (2003) J Chem Phys 119:224–231
121. Maroulis G (2008) J Chem Phys 129:044314
122. Maroulis G (2000) J Chem Phys 113:1813–1820
123. Rodriguez J, Laria D, Marceca EJ, Estrin DA (1999) J Chem Phys 110:9039–9047
124. Ghanty TK, Ghosh SK (2003) J Chem Phys 118:8547–8550
125. Maroulis G (2012) Int J Quant Chem 112:2231–2241
126. Frisch MJ, Pople JA, Del Bene JE (1985) J Phys Chem 89:3664–3669
127. Champagne B, Perpète EA, Jacquemin D, van Gisbergen SJA, Baerends EJ, Soubra-Ghaoui C, Robins KA, Kirtman B (2000) J Phys Chem A 104:4755–4763
128. Maroulis G (2012) Chem Phys Lett 525–526:49–53
129. Maroulis G, Haskopoulos A (2012) Comp Theor Chem 988:34–41

Struct Bond (2012) 149: 131–158
DOI: 10.1007/978-3-642-32753-7_4

Descriptors as Probes for Inter-Molecular Interactions and External Perturbation

Sourav Pal

Abstract In this article, different reactivity descriptors have been reviewed. Conceptual density functional theory is an important area where a lot of interesting developments in terms of descriptors and molecule interactions have taken place in recent years. This review will highlight our own work on such development of descriptors and their applications to characterise stability and reactivity in molecule systems.

Keywords Conceptual density functional theory · Global reactivity descriptors · Local reactivity descriptors · HSAB · Electric field effect · Local HSAB

Contents

S. Pal (✉)
CSIR-National Chemical Laboratory, Dr. Homi Bhabha Road, Pune 411 008, India
e-mail: s.pal@ncl.res.in

1 Introduction

In recent years, chemical reactivity theory has become an extremely important tool for characterising stability and reactivity of molecular species [1–6]. Density functional theory based reactivity descriptors have been developed in the recent years and applied to several problems relating to chemistry and biology [7–15]. Several conceptual developments were derived from density-based theories. Two broad types of descriptors, e.g. global reactivity descriptors (GRD) [6, 16–22] as well as local reactivity descriptors (LRD) [6, 23–26] have been developed. The first ones relate to the overall stability and reactivity of the system. The second class of descriptors relates to the local reactivity of the systems. These are potentially more useful to chemical and biological recognition processes. There have been significant developments of GRD as well as LRD in recent years. Excellent applications [7–12, 15, 27] of these to several interesting problems have also been made in the last several years. These have also provided a platform for conceptual understanding of the reactivity and stability of chemical systems.

GRDs are mainly based on the response of total energy with respect to the number of electrons. The most important of these are chemical potential and hardness. The latter, in particular, has been of great recent interest as a criterion of stability. A formal principle of maximum hardness (PMH) was proposed by Parr and Pearson [16, 19] and proved by Parr and co-workers [28]. Subsequently, this has been the subject of intense study and various verifications of this principle have been attempted. However, a more rigorous and systematic test of this principle was undertaken by Pal and co-workers [29], who, by the use of symmetric as well as asymmetric stretch of water molecules, established that the principle holds good under the conditions of constant chemical potential. The derivatives with respect to the number of electrons have been traditionally obtained by the finite field method. In the work by Pal et al. the calculations of the chemical potential and hardness were also carried out by the state-of-the-art coupled-cluster method [29]. A part of these findings was later proved by Makov [30] by group theoretic considerations. At the same time, Sebastian [31] showed that the PMH cannot hold good under general conditions and the conditions under which the principle holds well were clarified by Parr and co-workers [32] in a later paper. The inverse of chemical hardness, i.e. softness, was related to polarisability, which also indicated the reactivity of the system. A relation between softness with polarisability [33] was rigorously shown for atoms. An approximate relation between these two quantities was demonstrated numerically for molecules. There were several studies relating the change of hardness with reaction coordinates [34–37]. For example, change in hardness with bond length for diatomic molecules was studied and a monotonic nature of hardness was observed [38]. Since the chemical potential also changes monotonically in this case, it strengthens the necessity of constant conditions of chemical potential for PMH. The effect of electric field on the global descriptors has also been studied and Pal and co-workers [39, 40] also observed an interesting numerical relation of polarisability of N-electron system with dipole moments of the N-electron and

($N \pm 1$) electron systems. While the global descriptors have been important in determining the overall stability and reactivity of chemical systems, the development using the LRD has really made the field more useful [18, 41, 42]. The basic local descriptor is the change of local electron density with respect to the number of electrons or the change of chemical potential with respect to local external potentials. These have been known as Fukui functions, characterising the local reactivity in a system. Related to these are local softnesses, defined as change of electron density with respect to chemical potential. Of greater chemical interest has been the condensed versions of these Fukui functions or softnesses, which characterise the more reactive atomic sites [42]. Within a system, electrophilic, nucleophilic and radical reactive sites have been identified, depending on how the derivatives are calculated using numerical method. However, in a reaction involving two molecules, the reactive sites have been identified by the use of local hard-soft-acid-base (HSAB) principle, first proved by Gazquez and Mendez [43, 44]. The principle states that the reactive sites of the two systems have equal or comparable local softnesses. However, a quantitative version of the local HSAB principle by Gazquez co-workers allowed evaluation of interaction energies between two molecules using only the descriptors of individual reactivity systems [44]. This had a lot of potential in simplifying the molecular recognition problem. Inspired by local HSAB principle, several numerical studies have been performed by Geerlings and co-workers [45, 46], Pal and co-workers [47] and others [7, 48–50] to validate the principle. In particular, it was shown that the principle holds good for ionic interactions and was validated using interaction of zeolites with gas molecules [51]. Later, Pal and co-workers [52] extended the formulae to interaction taking place on multiple sites simultaneously, as in the case of DNA base pair interactions [53]. The latter development can be of interest in all cases, where bonding takes place simultaneously on different sites. Depending on the nature of bonding, two different types of formulae for interaction energies were proposed [53]. Parallel to these developments, a lot of attention has been paid to the method of calculation of Fukui functions.

It has been shown by Pal et al. [54, 55] that Hirshfeld population leads to intuitively correct non-negative condensed Fukui function. The effect of change of weightage of electron density on atoms on molecule in case of other population analysis due to the change of electron number by one may lead to negative Fukui function. Several studies involving condensed Fukui functions have shown negative values and this represents a serious problem in interpretation of results. Hirshfeld population analysis thus solves this problem. Pal and co-workers [56] have also discussed the separability at Fukui functions. For intra-molecular reactivity, relative condensed electrophilic and nucleophilic descriptors have been proposed by Roy et al. [57], which have proved to be extremely reliable. There have been attempts to define descriptors for other chemical phenomena. Tanwar et al. [58] have developed such a descriptor to correlate the hydrogen bond induced shift of vibrational frequency.

In this review article, we will highlight various facts of reactivity theory, developments of global and local descriptors, applicability of this descriptor stability and reactivity to various chemical and biological recognitions. Though a vast amount of work has been done by several workers, we will focus mostly on the work done by the author. The review is organised as follows: In the next section, we will review the development of global descriptors. Section 3 deals with various applications and insights originating for the use of global descriptors. Section 4 onwards presents the local descriptors and how these provide an important tool of reactivity.

2 Development of Global Descriptors

In density functional theory, the ground state energy of an atom or a molecule can be expressed as a unique functional of electron density due to the existence theorem proved by Hohenberg and Kohn [6, 59]:

$$E[\rho] = F_{\mathrm{HK}}[\rho] + \int \rho(r)v(r)\mathrm{d}r, \tag{1}$$

where $v(r)$ is the external potential and the universal Hohenberg–Kohn function $F_{\mathrm{HK}}[\rho]$ contains the kinetic energy, classical coulomb energy and exchange correlation energy. Density-based insight has given rise to several new qualitative concepts. One of the important parameters for a species in the electronegativity of Pauling [60] and Mulliken [61] is defined as follows:

$$\chi = \frac{1}{2}(I + A), \tag{2}$$

where I and A are the ionisation potential and electron affinity, respectively. This is a measure of the escaping tendency of electron from the species. The negative of this is defined as chemical potential. The original rationalisation of Eq. (2) was given by Mulliken [61] by considering electron transfer from A to B and from B to A for A and B of equal electronegativity. Since the energy associated with these electron transfers will be equal to $I_{\mathrm{A}} - A_{\mathrm{B}}$ for first case and $A_{\mathrm{A}} - I_{\mathrm{B}}$ for the second, Mullikan arrived at $(I + A)$ as a scale for electronegativity. Chemical potential $\mu(\equiv -\chi)$ has been defined as the first derivative of energy with respect to the number of electrons at constant external potential:

$$\mu = -\left(\frac{\partial E}{\partial N}\right)_v. \tag{3}$$

However, N is only an integral number; the calculation of chemical potential for atoms and molecules has not been very simple. The finite difference approximation [62] leads to

$$\mu = -\frac{1}{2}(I + A). \tag{4}$$

This fits in nicely with the definition of electronegativity for Eq. (2). Parr and Pearson [17] considered the next most important thing which was the second derivative of E vs. N curve and defined this as an operational definition of hardness (also called absolute hardness)

$$\eta = \frac{1}{2}\left(\frac{\partial^2 E}{\partial N^2}\right)_v. \tag{5}$$

This way hardness is defined as a quantity resulting in change or deformation of the system. Using finite difference formula, an operational definition of hardness was derived as

$$\eta = \frac{1}{2}(I - A). \tag{6}$$

One can clearly see that hardness is change of chemical potential with respect to the number of electrons and thus resists the change. Hardness is always non-negative and for bulk metals, it is zero. Softness S was defined as inverse of hardness:

$$S = \frac{1}{2\eta} = \frac{1}{I - A}. \tag{7}$$

Softness has been related to the overall reactivity of the system and hardness has been related to the stability. Using Koopmans' approximation, ionisation potential I can be associated with the negative of the energy of the highest occupied molecular orbital (HOMO) and the electron affinity is related to the energy of negative of the lowest unoccupied molecular orbital (LUMO). Thus, hardness η, under this approximation, [16, 63, 64] is

$$\eta = \frac{1}{2}(\varepsilon_{\text{lumo}} - \varepsilon_{\text{homo}}). \tag{8}$$

Fukui defined LOMO–HOMO gap as the stability of a molecular species and has used this as an indicator of the stability of organic systems, in particular.

This second derivative of energy with respect to the number of electrons plays an important part in the chemical reaction of two systems A and B with different electronegativities. Upon electron transfer from the more electronegative atom A to the less electronegative B, the electronegativity of B increases, while the electronegativity of A decreases. The decreasing nature of $\chi(\text{or } \mu)$ with the increase in the number of electrons was proposed by Huheey [65]. Thus, during chemical reaction, electron transfer takes place to equalise electronegativity known as Sanderson's electronegativity equalisation principle [66–70]. Using a Taylor series of chemical potential with respect to the number of electrons up to linear term, one can write for any system:

$$\mu = \mu^{\text{o}} + \eta^{\text{o}}(N - N_{\text{o}}), \tag{9}$$

where the electron number of the system changes from N_0 to N and μ^o and η^o are chemical potential and hardness of the individual system before the electron transfer. Using equalisation principle, one can arrive at an electron transfer ΔN

$$\Delta N = \frac{\chi_A^o - \chi_B^o}{\eta_A^o + \eta_B^o} = \frac{\mu_B^o - \mu_A^o}{\eta_A^o + \eta_B^o}. \tag{10}$$

This shows that harder the interacting partners, the lower the charge transfer. All these show the importance of hardness in the stability of the systems.

Formally, Pearson and Parr stated the principle of maximum hardness (PMH) [16, 19, 28, 63, 71] stating that at constant external potential v and chemical potential μ, the ensemble with the maximum hardness is the most stable. Though the condition of constant external potential and chemical potential severely restricted the scope of application of PMH, several workers calculated hardness using its operational definition [17] (Eq. (6)), using Koopmans' approximation [16, 63, 64] and tried to correlate stability with the calculated hardness. However, comparison across different systems was not so relevant. The most systematic test of PMH was carried out by Pal and co-workers [29], who studied H_2O molecule as an example. They studied both symmetric as well as asymmetric stretch of water molecule. The chemical potential and hardness were calculated by Eq. (6), where I and A were calculated by the most extensive *ab initio* coupled cluster theory. One valence Fock space multi-reference coupled cluster theory was used. Thus, this went far beyond the Koopmans' approximation normally used for such calculations. It was observed that for symmetric stretch, the chemical potential as well as hardness increased monotonically, thus showing no PMH. This was observed for equilibrium H–O–H bond angle as well as other values of this angle. On the other hand, it was observed that starting from any symmetric geometry, an asymmetric stretch led to generally invariant chemical potential. In this case, it was seen that the chemical hardness goes through a maximum effect at the symmetric point. This conclusively proves that for an ensemble of systems with the same chemical potential (for asymmetric stretch, one may assume that external potential on the electrons due to the nuclei is also roughly constant), there exists a maximum in chemical hardness at the locally symmetric point. Thus, many local maxima of chemical hardness exist. What was interesting was that various locally symmetric configurations were generated by changing bond angles and bond lengths such that the chemical potential of these are equal to the one of the equilibrium geometry of H_2O molecule. Comparing the hardnesses of the ensemble of these systems, it was indeed observed that the hardness of the equilibrium geometry was the maximum. This indeed was a more conclusive evidence of the PMH, though it showed that the constraint of equal chemical potential is an important one in the PMH. The results by Pal et al. were also obtained using a rigorous *ab initio* coupled-cluster calculation.

Hardness profile along the reaction coordinate also goes through a minimum at the transition point [72]. There are several other numerical calculations which confirm the maximum hardness principle [73–79]. The condition of constant chemical potential for small asymmetric vibrations about symmetric equilibrium position was also suggested by Pearson and Palke [80] and Chattaraj et al. [81]. Makov [30], using symmetry-based picture, predicted that energy, chemical potential and hardness will be extremum with respect to asymmetric variation about the symmetric point, thus partially explaining the results of Pal et al. [29], Pearson and Palke [80] and Dutta [72]. However, symmetry principles do not determine whether these are maxima or minima. Sebastian [31], around the same time, showed that the PMH cannot indeed be fulfilled in general. This prompted Parr and co-workers to redefine the conditions of PMH appropriately. The constancy of chemical potential indeed came out as one of the important conditions.

3 Relation of Hardness and Softness with Other Molecular Properties

While chemical hardness attracted a lot of attraction due to PMH, it was interesting to note at the same time there is a lot of interest to correlate hardness and softness with other properties, in particular, polarisability. For atoms, it was well known that polarisability (α) was proportional to the volume of the atom and hence $\alpha^{1/3}$ bears a linear relation with the atomic radius r [82, 83]. It has been observed that softness increases linearly with r and hence the linear relation of $\alpha^{1/3}$ and softness is rigorously expected for atoms. For molecules, however, the charge distribution is nonspherical and the volume of a molecule does not necessarily scale with the cubic power of any parameter. However, Roy et al. [33] numerically studied the change of hardness (η) with $\alpha^{1/3}$ for some simple molecular systems. They chose a series of molecules CH_4, CH_3F and CH_3Cl and calculated α and (η) with bond stretch of C–H, C–F, and C–Cl respectively. They found that the $\alpha^{1/3}$ bears a dominant linear relationship with hardness (η). $\alpha^{1/3}$ decreases with hardness and the α itself was defined as an arithmetic average of the diagonal part of the polarisability tensor. For the sake of calculation of α, values were obtained in FF approach and at SCF level using Huzinaga-Dunning double zeta basis set. Similar studies were done by several other authors [84–93]. Ghosh et al. [92, 94–96] showed that a rigorous relation can be derived for atoms; the same is not possible in molecular cases. However, the general principle of increase of S with α is validated by the PMH principle. Greater S or less η has been associated with the increase of α and hence, reactivity of the system. Softness has been associated with the global reactivity of the system [97] and η has been associated with the stability of the system. For chemistry, however, what is important is to find a region or atoms in the molecule which may be reactive. This cannot be explained by GRD, referred so far in this article. The development of LRD, however, has led to a remarkable progress in examining the locally reactive points in a molecule during its interaction with another molecule. There has been a greater interest in recent years and hence, this development of LRD as explained in the next section onwards is of great relevance to general molecular interaction processes.

4 Local Reactivity Descriptors

4.1 The Need of Local Reactivity Descriptors

The mathematical formulation and definition of GRDs provide information about the overall properties of the chemical systems and are also useful for analysing the thermodynamic aspects of chemical reactions. However, the GRDs cannot identify the reactive part or site of the molecule. Most of the chemical reactions are primarily concerned with the atoms, group of atoms or a specific site in a molecule. The important issues relate to the effect of charge or density fluctuations in chemical reactivity and the observed reactivity trends. The chemical reactivity for a system is determined by the sensitivity of electron density to the perturbations. Therefore, an appropriate definition was required in addition to the global descriptors [18]. A basic local descriptor is related to the change of the local electron density to the chemical potential of the system. This is known as local softness.

The local softness $s(r)$ of a chemical species or molecule is given by [90]

$$s(r) = \left(\frac{\partial \rho(r)}{\partial \mu}\right)_{v(r)}, \tag{11}$$

such that,

$$\int s(r)\,\mathrm{d}r = S, \tag{12}$$

where

$$S = \left(\frac{\partial N}{\partial \mu}\right)_{v(r)}. \tag{13}$$

Further, local softness can be written as

$$s(r) = \left(\frac{\partial \rho(r)}{\partial N}\right)_{v(r)} \left(\frac{\partial N}{\partial \mu}\right)_{v(r)} \tag{14}$$

$$= f(r)S, \tag{15}$$

where $f(r)$ is defined as the Fukui function [91]. Fukui function is often described to be the most basic local descriptor. It is related to the change of electron density with respect to the number of electrons. The overall integrated Fukui function leads to 1.

The Fukui function contains relative information about reaction to the different regions in a given molecule. In contrast, the local softness integrates to the total value of softness (S), which is different for different molecules. Thus, when comparison of reactivities across different molecules is warranted, local softness has been the preferred choice. Fukui function has also been interpreted as change in the chemical potential with respect to local external potential.

$$f(r) = \left(\frac{\partial \mu}{\partial v(r)}\right)_N. \tag{16}$$

The above equation arises because density, $\rho(r)$ itself in the first partial derivative of energy E with respect to total number of electrons N at constant external potential, $v(r)$. The derivative $\partial\rho(r)/\partial N$ at constant $v(r)$ has been also interpreted by Parr and Yang [98, 99] as the density of HOMO and LUMO. The partial derivative is difficult to evaluate because of N-discontinuity problem in atom and molecules. A finite field definition of change of electron density from N-electron to $(N \pm 1)$-electron systems is the preferred choice of evaluation of $f(r)$. This introduces the left and the right derivative of density with respect to electrons which has been interpreted as electrophilic $f^+(r)$ and nucleophilic $f^-(r)$, respectively:

$$f^+(r) = \left(\frac{\partial \rho(r)}{\partial N}\right)^+_{v(r)}, \tag{17}$$

$$f^-(r) = \left(\frac{\partial \rho(r)}{\partial N}\right)^-_{v(r)}. \tag{18}$$

$f^+(r)$ signifies an attack by a nucleophile leading to an electron increase in a system; similarly, nucleophilic Fukui function $f^-(r)$ describes an attack by an electrophile leading to a electron decrease in a system. The larger the $f^+(r)$, the greater the possibility of nucleophilic attack in that region of the system, and conversely, the larger the $f^-(r)$, the greater the possibility of electrophile attacking the system. In a similar vein, one can also define Fukui function, $f^o(r)$ which describes the probability of the radical attack of the system which is the arithmetic average of $f^+(r)$ and $f^-(r)$.

To describe a size selectivity of atom and molecule, it is, however, necessary to describe the reactivity not in terms of region of space in a molecule but atom in a molecule. This warrants condensation of the two basic descriptors $f(r)$ and $s(r)$, around each atomic site. This condensation requires a measurement of the population of electrons in each atom in molecule, which is derivable from the overall density of molecule, $\rho(r)$. Assuming $q_k^{N_0}$ as the electron population of an atom k in a molecule, with N_o number of electrons, one can define three different types of condense Fukui function as proposed by Yang and Mortier [42] which precedes Nalewajski's [100] atoms in molecule approach.

$$f_k^+ \approx q_k^{N_0+1} - q_k^{N_0}, \qquad \text{nucleophilic attack,} \tag{19a}$$

$$f_k^{\ -} \approx q_k^{N_0} - q_k^{N_0 - 1}, \qquad \text{electrophilic attack,} \tag{19b}$$

$$f_k^{\ \text{o}} \approx \frac{1}{2}\left(q_k^{N_0+1} - q_k^{N_0-1}\right), \qquad \text{radical attack,} \tag{19c}$$

where $q_k^{N_0+1}$ and $q_k^{N_0-1}$ are the electron population of the atom k in the molecule. The atom having more condense f_k^+ ,f_k^- andf_k^0 will be more prone to nucleophilic, electrophilic and radical attack, respectively. The corresponding local softness can be defined as

$$s_k^+ = f_k^+ S, \tag{20}$$

$$s_k^- = f_k^- S, \tag{21}$$

and

$$s_k^0 = f_k^0 S. \tag{22}$$

The larger Fukui function describes to a larger reactivity of an atom by Parr and Yang [18, 41]. Later Ayer and Parr [101] showed that a Fukui function can also be defined by minimising hardness functional, $\eta(\rho_{N_0}, \triangle\rho_{+1})$, where $\triangle\rho_{+1}$ signifies the density of extra electron, which implies that it integrates to 1.

4.2 The Sign of Fukui Function Values

Although no formal proof has been proposed regarding the sign of Fukui function indices, it is intuitively assumed that the Fukui function will be positive. If the electron number of the system increases incrementally, it is assumed that the electron density should not be depleted from any particular site (or atom, in case of condensed definition) or any region of the molecule. On the other hand, it has been observed in many cases that the condensed Fukui function has negative values. This is especially true using some of the more standard population Mulliken population analysis and Löwdin population analysis to calculate electron population on an atom. The problem has been investigated by Roy, Pal and Hirao [54, 55, 102] in great detail.

Equation (16) can be rewritten as:

$$f(\bar{r}) = \underset{\Delta N \to 0}{\text{Lt}} \frac{\rho_{N+\Delta N(\bar{r})} - \rho_{N(\bar{r})}}{\Delta N}. \tag{23}$$

Equation (23) implies the analytic definition of Fukui function is valid at the limit of infinitesimally small ΔN. However, to define the condense Fukui function or local softness, we allow numerical change of electron number by +1 or −1. This leads to a significant change due to the large relaxation in corresponding cations and anions. Thus, relaxation may cause depletion or accumulation of electron density from (or in) a region, even though the global number of electron increases or decreases by 1.

The same may lead to negative Fukui function and subsequently, negative condense Fukui function using some arbitrary population analysis. However, the condense Fukui function values are extremely sensitive in both the level of calculation; in addition to the partitioning scheme to obtain the electron population, various partitioning schemes are available, e.g. Mulliken [103], CHELPG [104], natural population analysis (NPA) [105], molecular electrostatic potential (MESP) [106], atom polar tensor based formalism [107–109], Bader's atom in molecules [110, 111] picture and Hirshfeld population analysis (HPA) [112]. Each of these population analyses is arbitrary, though, Mulliken population analysis (MPA) and Löwdin population analysis (LPA) are often the choice of partitioning scheme. Numerically, it is seen that both MPA and LPA [113, 114] partitioning scheme leads to negative Fukui function rendering it difficult to rank order the reactivity of atoms in the molecule. Each of the population analysis, particularly, MPA and LPA may have its own value of electron population for $(N \pm 1)$ electron system. Each of the methods, however, is sensitive to basis set and can give unreasonable physical picture for charge distribution. For example, MPA over emphasises the ionic character of the molecule. The molecular electrostatic potential (MESP) and natural population analysis (NPA) are also arbitrary population analysis and can still lead to negative FF. The MESP is based on the electrostatic potential for a unit positive charge at point r in the vicinity of the given molecule:

$$V(r) = \sum_A \frac{Z_A}{|r - R_A|} - \sum_{mn} P_{mn} \int \frac{\phi_m \phi_n}{|r - r'|} \mathrm{d}r', \tag{24}$$

where P_{mn} refers to a particular density matrix element obtained from the SCF-MO calculation; are the atomic basis wavefunctions. The first term is summed over all nuclei, and the second term is the contribution of all electrons in the molecule. NPA often exhibits numerical stability with reference to MPA and describes the compound of high ionic character more reliably. It is based on the population derived from orthonormal natural orbitals. The natural population on this basis, n_i is the occupancies of the electron in the natural orbitals and vigorously satisfies the Pauli exclusion principle, $0 \leq n_i \leq 2$. The population of electrons in an atom (A), n_A is the sum of the population of the natural atomic orbital centre of atom A:

$$n(A) = \sum_A n_i(A). \tag{25}$$

NPA can particularly resolve the basis set sensitivity of MPA. However, it still leads to negative Fukui function. Roy et al. [54, 55] showed, on the other hand, that the Hirshfeld population analysis (HPA) can generate non-negative Fukui function and presented the argument in favour of this.

HPA [112, 115] defines atomic charges by dividing density between atoms in the molecule in the form of ratio of the pro-molecular density, i.e. the density before the molecule formed. This charge in the density with the molecule formation is often known as $\rho_d(\bar{r})$ and is the difference between the molecular and pro-molecular density:

$$\rho_d(\bar{r}) = \rho^{\text{mol}}(\bar{r}) - \rho^{\text{pro}}(\bar{r}) = \rho^{\text{mol}}(\bar{r}) - \sum_{\alpha} \rho_\alpha(\bar{r} - \bar{R}_\alpha), \tag{26}$$

$$q_\alpha = -\int \rho_d(\bar{r})\omega_\alpha(\bar{r})d^3\bar{r}. \tag{27}$$

$\rho^{\text{mol}}(\bar{r})$ is the molecular density at site $\bar{r}$, $\rho^{\text{pro}}(\bar{r})$ is the pro-molecular density at site $\bar{r}$ and $\rho_\alpha(\bar{r} - \overline{R_\alpha})$ is the ground state electron density centred at $\overline{R_\alpha}$. The effective electrons q_α on an atom α can be computed from the deformation density, using Eq. (27), where $\omega_\alpha(\bar{r})$ is the weight function, the relative weightage of atom α in the pro-molecular density at the position $\bar{r}$.

$$\omega_\alpha = \rho_\alpha(\bar{r} - \bar{R}_\alpha)\left[\sum_{\beta} \rho_\beta(\bar{r} - \bar{R}_\beta)\right]^{-1}. \tag{28}$$

Thus, in HPA, it is assumed that the molecular charge density at each point is divided among the atoms of the molecule in proportion to their respective contribution before formation of the molecule. This is often called "stock-holder partition" where each atoms gain or lose in direct proportion of its share in the capital investment (in the pro-molecular formation). The HPA gives the results of electron population in a molecule in contradiction to our intuition and often the charges obtained by the HPA are not correct to a given N-electron system. Ideally, Hirshfeld charges would be determined primarily by the electron donating or electron withdrawing strength of an atom relative to its neighbouring atom. Intuitively, however, when HPA used to calculate the condense Fukui function may lead to non-negative condensed Fukui function. f_k^+ and f_k^- can now be defined as

$$f_k^+ = \int \Delta\rho^+(\bar{r})\varpi_k(\bar{r})d^3\bar{r} \qquad \text{for nucleophilic attack,} \tag{29a}$$

$$f_k^- = \int \Delta\rho^-(\bar{r})\varpi_k(\bar{r})d^3\bar{r} \qquad \text{for electrophilic attack,} \tag{29b}$$

where

$$\Delta\rho^{+}(\bar{r}) = \rho^{\text{mol}}_{N_0+1}(\bar{r}) - \rho^{\text{mol}}_{N_0}(\bar{r}), \tag{30a}$$

$$\Delta\rho^{-}(\bar{r}) = \rho^{\text{mol}}_{N_0}(\bar{r}) - \rho^{\text{mol}}_{N_0-1}(\bar{r}). \tag{30b}$$

In Eqs. (30a) and (30b), $\rho^{\text{mol}}_{N_0}(\bar{r}), \rho^{\text{mol}}_{N_0+1}(\bar{r})$ and $\rho^{\text{mol}}_{N_0-1}(\bar{r})$, denote the molecular electron densities at position $\bar{r}$ for $N_o, N_o + 1$ and $N_o - 1$, respectively. The expression of the condense Fukui function because of common weight factor $\omega_k(\bar{r})$ in N_o, $N_o \pm 1$, electron systems may impact to believe the non-negative condense Fukui function. The actual numerical derivation of Eq. (23), the ΔN, should be infinitesimally small. This implies that the weight factor of N_o and $N_o \pm 1$ electron system should be nearly the same, since ΔN is infinitesimally small. Thus, though, HPA is artificially sound for N_o and $N_o \pm 1$ electron system, it is a way to cancel the error of numerical derivative obtained using $\Delta N = \pm 1$. In MPA and other population analyses, this weight factor is not common, e.g. in MPA and LPA the weight factors would depend on SCF AO to MO coefficients which are different from $N_o, N_o + 1$ and $N_o - 1$ electron system. Hence, the negative Fukui function can be ascribed to the partitioning scheme and not to the N-discontinuity problem [102]. Nalewajski and Parr [100] showed that the maximum conservation of information contains of isolated atoms upon molecular formation are best recovered by Stock holder partitioning of electron density.

4.3 Interpretation and Applications of Fukui Functions

A significant number of studies have been done on the effect of atomic orbital basis set. Geerlings, in his result of QCISD [116, 117], showed using Bader population analysis that B3LYP and especially B3PW91 perform better than HF and MP2. Arumozhiraja and Kolandaivel [118] also investigated that condense Fukui function is highly sensitive to the basis set and population analysis. Among the few studies of condensed Fukui function on open shell system, Mishra and Sannigrahi [119, 120] have noticed that the spin contamination effect is negligible in finite difference approach. Chandra and Nguyen [7] first studied the radical attack on non-radical system using Fukui function. An extension of Fukui function concept to a reactant description of the chemical reaction was done by Nalewajski and Korchowiec [121].

Parr [122, 123] has shown that the Fukui function can be expressed in terms of Lowest Unoccupied molecular orbital (LUMO) and Highest Occupied Molecular Orbital (HOMO) as follows:

$$f^{+}(r) = \left|\phi(r)^{\text{LUMO}}\right|^{2} = \rho(r)^{\text{LUMO}}, \tag{31a}$$

$$f^{-}(r) = \left|\phi(r)^{\text{HOMO}}\right|^{2} = \rho(r)^{\text{HOMO}}. \tag{31b}$$

It is clear from Eqs. (31a) and (31b) that $f(r)$ is the DFT analogue of the frontier orbital regioselectivity for nucleophilic $f^{+}(r)$ and electrophilic $f^{-}(r)$ attack and it is a restatement of frontier molecular orbital (FMO) theory. However, the last statement may not be true because Fukui function includes the effect of electron correlation and orbital relaxation that are a priori neglected in FMO theory. Flurchik and Bartorotti [124] noticed considerable differences between HOMO/LUMO density and Fukui function value. Chattaraj et al. [125] and Pacios et al. [126, 127] proposed gradient approximation for Fukui function to avoid the calculation of metastable anion to calculate Fukui function using finite difference approximation:

$$f(r) = \frac{\rho(r)}{N}\left[1 + \alpha\phi\left(r, \rho(r), \nabla\rho, \nabla^{2}\rho, \ldots\right)\right]. \tag{32}$$

This can also be written as

$$f(r) = \frac{\rho(r)}{N} + \frac{\alpha}{N}\rho_{0}^{-2/3}\left\{\left[\left(\frac{\rho}{\rho_{0}}\right)^{2/3} - 1\right]\nabla^{2}\rho - \frac{2}{3}\left(\frac{\rho_{0}}{\rho}\right)^{2/3}\frac{\nabla\rho_{0}.\nabla\rho}{\rho}\right\}, \tag{33}$$

where ρ_0 which determines the parameter α is the density of the nucleus. The first formulation of varional principle of Fukui function has been done by Chattaraj et al. [128] and was later implemented by De Proft et al. [129]. According to Nalewajski's work, Fukui function can be obtained from single Kohn sham calculations [130],

$$f(r) = f^{\text{F}}(r) + f^{\text{R}}(r). \tag{34}$$

Here, the second term $f^{\text{R}}(r)$, corresponding to the orbital relaxation, was missing in frontier orbital theory and the first term $f^{\text{F}}(r)$ is the frontier orbital term.

4.4 Relative Electrophilicity $\left(s_k^{+}/s_k^{-}\right)$ and Relative Nucleophilicity $\left(s_k^{-}/s_k^{+}\right)$

The condensed local softness (or Fukui function) is very successful in explaining the experimentally observed intramolecular reactivity trends. However, it has been observed by Roy et al. [57] that s_k^{+} and s_k^{-} fail to reproduce experimentally observed intramolecular trends in aliphatic and aromatic carbonyl compounds in some cases. To obtain the correct reactivity trends of those molecules, they proposed new reactivity descriptors [131]: Relative Electrophilicity $\left(s_k^{+}/s_k^{-}\right)$ and Relative Nucleophilicity

$\left(s_k^-/s_k^+\right)$. Due to the ratio of s_k^+ and s_k^- or vice versa, the new two reactivity descriptors should be less sensitive to the basis set and correlation effects which are more prominent of the individual s_k^+ and s_k^-. So, it was prescribed that the site having highest value of $\left(s_k^+/s_k^-\right)$ is the most probable site for nucleophile attack and the site having highest value of $\left(s_k^-/s_k^+\right)$ is the most probable site for electrophile attack.

To predict the site selectivity of aliphatic and aromatic carbonyl compounds and sites of gas phase protonation of aniline and substituted aniline, this newly defined reactivity descriptor gives better description. However, it fails to give the correct rank ordering of the reactive site of molecules when it attains negative value. Although individual s_k^+ and s_k^- values suggested the preferable site towards nucleophilic and electrophilic attack, their ratio offers contradictory rank ordering. This indirectly raises the question whether negative FF indices are a physical reality or is it an artefact of the approximation and methodology adopted to evaluate it in a popular usable form.

4.5 Group Softness

Some of the chemical reactions occurred by participation of group of atoms rather than one atom. Krishnamurty et al. [132] explained those kinds of reaction using the concept of group softness. Group softness is the sum of softness values of the group of relevant atoms and it can be written as

$$s_g = \sum_{k=1}^{n} s_k \;\; s_g = \sum_{k=1}^{n} s_k, \tag{35}$$

where n is the number of atoms bonded to the reactive atom, s_k is the local softness of the atom k and s_g is the group softness obtained after summing the local softness of all the neighbouring atoms. In a similar fashion, later Shetty et al. [133] defined group FF f_g which is the sum of the FF of the atoms of the group and obviously obeyed the relation $f_g = \left(s_g/S\right)$, where S is the global softness of the systems. In the case of intramolecular reactivity, f_g and s_g play the same role, but on the other hand to compare reactivity across different molecules f_g and s_g will not necessarily give the same trend because global softness of the system varies.

4.6 Separability of Local Reactivity Descriptors

Separability is a very important term in quantum mechanics. When N number of monomers is creating a quantum state without any quantum correlation between the sub-systems, the state is called perfectly separable. In a physical sense, the energy of such system will be N times the energy of the monomer. If any system following this property is said to be size consistent [113, 134, 135]. It has very important role in electronic structure calculations. It helps to calculate the energy of a system which has been broken into many fragments. The computation methods must be size consistent to describe the breakup of the system into fragments. Hartree–Fock, full configuration interactions (CI), Perturbation theory and coupled-cluster methods are size-consistent methods while truncated CI is not size consistent [113, 136]. Although it has been nicely understood that total energies of non-interacting fragments are the summation of the energy of each fragments and the electronic densities of the fragments should separate out, it is not clearly understood what happens to the Local Reactivity (LRDs) [137]. Pal and co-workers [38, 56] evaluated the global parameter, chemical potential and hardness at the non-interacting limit in addition to the separability of LRDs. Significant effort has been made by Pal and co-workers [56] to derive the separability of Fukui function, taking two different types of systems, viz. AB and dimer A_2. They considered a complex AB which dissociates into two fragments A and B. Electron density of such system can be written as a sum of the electron density of A and B:

$$\rho^{\mathrm{AB}}(r) = \rho^{\mathrm{A}}(r) + \rho^{\mathrm{B}}(r), \tag{36}$$

where $\rho^{\mathrm{AB}}(r)$ is the electron density of the system AB at point r. $\rho^{A}(r)$ is the density at r due to fragments A. Similarly, $\rho^{B}(r)$ is the electron density at point r due to fragments B. By recalling the definition of Fukui function one can write:

$$f_{\mathrm{AB}}(r) = \left(\frac{\partial\rho^{\mathrm{AB}}(r)}{\partial N}\right)_{v(r)} = \left(\frac{\partial\rho^{\mathrm{A}}(r)}{\partial N}\right)_{v(r)} + \left(\frac{\partial\rho^{\mathrm{B}}(r)}{\partial N}\right)_{v(r)} \neq f_{\mathrm{A}}(r) + f_{\mathrm{B}}(r). \tag{37}$$

In Eq. (37), $N = N_{\mathrm{A}} + N_{\mathrm{B}}$, is the total number of system AB, where N_{A} is number of electrons in A and N_{B} is number of electrons in B. The inequality in the above equation signifies that the FF of complex AB cannot be expressed as a sum of FF of the isolated fragments and hence, FF is not size extensive. In that context, Pal and co-workers [56] also showed, in case of complex dissociation, the FF of the complex reduces to the FF of individual fragment depending on whose cation or anion is more stable. Therefore, the following expression can be written for FF:

$$\begin{aligned} f^{+}_{\mathrm{A}_k,\mathrm{AB}} &= f^{+}_{\mathrm{A}_K}, \\ f^{+}_{\mathrm{B}_k,\mathrm{AB}} &= 0, \\ f^{-}_{\mathrm{A}_k,\mathrm{AB}} &= 0, \\ f^{-}_{\mathrm{B}_k,\mathrm{AB}} &= f^{-}_{\mathrm{B}_K}. \end{aligned} \tag{38}$$

Equation (38) implies that when the complex dissociates, the condensed electrophilic FF of the atoms of the fragment for which anion is formed goes towards the condensed electrophilic FF of the isolated fragment, while for the other atoms it goes towards zero. The condensed nucleophilic FF of the atoms of the fragment for which cation is formed goes towards the condensed nucleophilic FF of isolated fragment, while for the atoms of the other fragment these values go to zero. In the case of the radical attack equation can be written as

$$\begin{aligned} f^a_{A_i,A2} &= f^a_{A_i,A}/2 \forall a = +,-,0. \\ f^0_{B_k,AB} &= f^-_{B_K}/2 \neq f^0_{B_k}. \end{aligned} \tag{39}$$

Equation (39) describes condensed FFs for radical attack.

However, for the A_2 type dimer FF was defined as

$$f^a_{A_i,A2} = f^a_{A_i,A}/2 \forall a = +,-,0 \text{ and } i = 1,2\ldots k. \tag{40}$$

Interestingly, Eq. (40) tells us FF of non-interacting dimer, at a point r, is the sum of the half of FF of the monomer, and the atom condensed FFs of the atoms of the non-interacting dimer would be half as that of the monomer. In a similar fashion, for non-interacting systems, the condensed FF of ith atom in a system A is n times less than that of the atom k in isolated A itself. Apart from condensed FFs, Pal and co-workers also defined other LRDs separability such as Relative Electrophilicity (RE), Relative Nucleophilicity (RN) and local softness [56]. The other LRDs also showed the same behaviour of separability in dimer A_2 type of systems.

4.7 *Effect of Perturbation on Reactivity Descriptors and HSAB Principle*

In this section, we briefly review the effect of perturbation on the descriptors and the HSAB principle, which can present a reliable estimate of the interaction energies using only the local descriptors of the reacting sites. Considerable work has been done on this subject and this, by itself, merits a separate article. However, due to constraint in space, we will describe these briefly here.

The effect of external perturbation on the local reactivity descriptors has been described by Senet, Fuentealba and co-workers [138–140] using perturbative analysis of energy and electron density with respect to the number of electrons and external field. Since $\rho(r) = \partial E/\partial v_{\text{ext}}$, response of $f(r)$ with N provides the same results as response of η with respect to v_{ext}. Perturbations have an impact on some specific strong, weak or Vander Waals bond type interactions due to redistribution of electron density of the interacting systems. Effect of homogeneous and inhomogeneous electric fields perturbation on a particular system has been studied by many theoretical chemists [141–144]. These effects are especially important in ordered crystalline environments such as solid oxides (e.g. zeolites and other metal oxides) and biological macromolecules [145–148]. Stability of such complexes highly depends on the perturbation. Hence, the prediction of the reactivity and stability of molecular systems in the presence of such environmental factors (solvent, electric field or point charges) becomes an important issue. Lot of effort has been given to study reactivity and stability of different complexes due to applied electric field in terms of the reactivity descriptors. Fuentealba and Cedillo [138] have derived an expression for the variation of Kohn-Sham Fukui functions under the external fields using energy-density perturbation methods. Kar et al. [39] studied the effect of electric field on the global reactivity descriptors (GRD) and local reactivity descriptors (LRD) of some molecular systems. The authors have noticed that a small amount of electric field is sufficient to distort the electron density distribution and hence gives a great impact on the reactivity of the systems as well [39, 149]. Another study by Kar et al. [149] described the change in descriptors and reactivity on non-linear molecules and observed, in the case of application of electric field perpendicular to the principal axis, there is a change in both GRD and LRD unlike linear molecules. By the help of their results again it has been proved that the variation of the hardness parameter is actually dependent on the net cooperative effect exhibited by the collection of all the atoms present in the molecule.

4.7.1 HSAB Principle

In 1963, Pearson [21] proposed hard soft acid base principle which states that hard acid prefers to combine with hard base and soft acid prefers to combine with soft base. There is an extra stabilisation for hard–hard and soft–soft combination. HSAB principle has been very successful at a qualitative level in rationalising most of the acid–base reaction. It has become very popular among the community of chemists because of its wider applicability and simplicity. However, theoretically quantification of qualitative principle is a great challenge to us. Many groups are working on this issue for explaining the relative bond strengths of acid–base complex, calculation of interaction energy between interacting molecules, etc. One of the earlier proofs of HSAB principle proposed by Parr and Pearson [17] was derived on the basis of energy perturbation with respect to the number of electrons. Following Parr's work, Nalewajski [150] employed both the number of electrons and external potential as the perturbation variables. Later, Parr and his collaborators [71] approached the proof of HSAB principle in a different way and this method was based on the minimisation of the interaction energy expression with respect to the softness parameter keeping other factors such as chemical potential and softness of the other reactants as constants. Recent study by Li and Evans [151] using the softness kernels and other local reactivity descriptors in their proof, brought out many interesting relationships between the HSAB principle and the Fukui's FMO theory. First, the theoretical proof for the HSAB principle using the Taylor series energy expansion in terms of the number of electron (N) as a perturbation variable came out in 1983 by Parr and Pearson [17]. According to them, the interaction energy can be expressed as

$$\Delta E = \frac{(\mu_{\mathrm{B}} - \mu_{\mathrm{A}})^2}{4(\eta_{\mathrm{A}} + \eta_{\mathrm{B}})}. \tag{41}$$

It can be observed from the Eq. (41), the differences in electronegativity or chemical potential drive the electron transfer and lower the energy of the interacting systems. Electron transfer process is assumed to take place continuously until the equilibrium is attained and it is referred to as the chemical potential or electronegativity equalisation process. In case soft acid and base $(\eta_A + \eta_B)$ is a small number, and for a reasonable difference in electronegativity, ΔE is substantial and stabilising. This explains the HSAB principle, in part: soft prefers soft but it does not explain the hard–hard preference (large denominator). Following this work, Nalewajski [150] pointed out that the hard–hard interaction can be described by the inclusion of the first-order contribution due to the perturbing external potential (i.e. $E(N,Z)$). The basic expression is written as

$$\Delta E_{AB} = \frac{\left(\mu_B^0 - \mu_A^0\right)^2}{2(\eta_A + \eta_B)} + \frac{\left(\mu_B^0 - \mu_A^0\right)(\alpha_B \Delta Z_B - \alpha_A \Delta Z_A)}{(\eta_A + \eta_B)} + \left(v_A^0 \Delta Z_A + v_B^0 \Delta Z_B\right), \quad (42)$$

where v is the electron-nuclear attraction per unit charge,

$$v = \frac{V_{n\varepsilon}}{Z} < 0,$$

$$\alpha = \frac{1}{2}\left(\frac{\partial \mu}{\partial Z}\right)_N = \frac{1}{2}\left(\frac{\partial v}{\partial N}\right)_Z < 0,$$

and $\beta = \frac{1}{2}\left(\frac{\partial v}{\partial Z}\right)_N < 0.$

In the context of HSAB principle, Chattaraj et al. [71] have derived an expression and made a statement, among the potential partners of a given electronegativity, hard likes hard and soft likes soft.

An alternative proof HSAB principle has been derived by Gazquez [152]. He has given an approximate total energy expression in terms of the chemical potential μ and the hardness η,

$$E[\rho] = \mu N_e - \frac{1}{2} N_e^2 \eta + E_{core}[\rho] \quad (43)$$

Here, N_e is an effective number of valence electrons and $E_{core}[\rho]$ denotes the core contribution to the total energy. Using the above expression, the authors showed that the interaction between species whose softness is approximately equal is energetically favoured, relative to the interaction between species whose softness is very different from each other. At the same time, Li and Evans [151] also demonstrated the HSAB principle, which is very similar to the Nalewajski's work. However, Li and Evans have included local descriptors and softness kernel in their proof rather than using the global descriptors alone. One of the significant features of their study is that the Fukui function is shown to be one of the key concepts in relating the frontier molecular orbital theory and the HSAB principle.

The final expression for the interaction energy, considering the interaction between two molecules *i* and *j* takes the form:

$$\begin{aligned}
\Delta E_{ij} = {} & \left(\mu_i - \mu_j\right)\Delta N + \left(\eta_i - \eta_j\right)^2 \Delta N^2 + \int \left[\rho_i(r)\Delta v_i(r) + \rho_j(r)\Delta \dot{v}_j(r)\right] \mathrm{d}r \\
& + \Delta N \int [f_i(r)\Delta v_i(r) - f_i(r)\Delta v_i(r)] \mathrm{d}r + \frac{\eta_i a}{2}\left(\int s_i(r)\Delta v_i(r)\mathrm{d}r\right)^2 \\
& + \frac{\eta_j a}{2}\left(\int s_j(r)\Delta v_j(r)\mathrm{d}r\right)^2 \\
& - \frac{a}{2}\int \left[s_i(r)(\Delta v_i(r))^2 + s_j(r)\left(\Delta v_j(r)\right)^2\right] \mathrm{d}r,
\end{aligned} \tag{44}$$

where *a* and *b* are constants and μ, η and *s* are the chemical potential, hardness, local softness, respectively.

There have also been many numerical studies that have supported the HSAB principle [153–161]. Single site based intermolecular interactions have been studied using Local HSAB Principle and interaction energy can be calculated by the expression [43, 44, 51, 77, 162]

$$\Delta E_{\mathrm{int}} = -\frac{1}{2}\left(\frac{\left(\mu_{\mathrm{A}} - \mu_{\mathrm{B}}\right)^2}{\eta_{\mathrm{A}} + \eta_{\mathrm{B}}}\right)_v - \frac{1}{2}N_{\mathrm{AB}}^2\left(\eta_{\mathrm{AB}} - \eta_{\mathrm{AB}}^*\right)_\mu, \tag{45}$$

where A and B are two interacting model systems, η_{AB} and η_{AB}^* are the hardness of the complex at the equilibrium and at the isolated state, respectively. Pal et al. [52, 53] have logically extended single-site local HSAB principle to multiple sites by assuming that the interaction occurs in a decoupled manner and the additivity of energy. In terms of the global softness of the systems A and B, Eq. (45) can be written as

$$\Delta E_{\mathrm{int}} \approx -\frac{\left(\mu_{\mathrm{B}} - \mu_{\mathrm{A}}\right)^2}{2}\left(\frac{S_{\mathrm{A}}S_{\mathrm{B}}}{S_{\mathrm{A}} + S_{\mathrm{B}}}\right)_v - \frac{\lambda}{4}\left(\frac{1}{S_{\mathrm{A}} + S_{\mathrm{B}}}\right)_\mu. \tag{46}$$

Equation (46) contains an ad hoc term λ which cannot be computed rigorously only through the softness of the molecular complexes. In the literature, there are several different definitions of this ad hoc parameter. In the initial study on the regioselectivity of enolate alkylation, Gazquez and Mendez [44] have used 0.5 for the value of λ without further justification. Geerlings and co-workers [163] also used the value of 0.5 and 1.0 for certain organic reactions. In the study by Mendez et al. [164] on 1,3 dipolar cycloaddition reactions, the dependence of the total interaction energy has been calculated at a local (dipole)–global (dipolarophile) level. Value of λ indicates that regioselectivity in the reactions between benzonitrile oxide and vinyl *p*-nitrobenzoate and 1-acetylvinyl *p*-nitrobenzoate is predicted correctly as long as $\lambda > 0.2$. Pal and co-workers [51, 165] have given an alternative definition of λ and justified for weak to moderately weak interaction. They have mentioned that the term λ can be written as the difference of electron densities of the system A before and after the interaction:

$$\lambda_{\mathrm{A}} = \sum_{i=1}^{p} \rho_{\mathrm{A}_i}^{\mathrm{eq}} - \sum_{i=1}^{p} \rho_{\mathrm{A}_i}^{0}. \tag{47a}$$

Alternately, the term λ can be defined as the difference of electron densities for the system B:

$$\lambda_{\mathrm{B}} = \sum_{j=1}^{q} \rho_{\mathrm{A}_j}^{\mathrm{eq}} - \sum_{j=1}^{q} \rho_{\mathrm{A}_j}^{0}, \tag{47b}$$

where the first terms of the right-hand side of the Eqs. (47a) and (47b) refer to the sum of the electron densities of each atom in A and B in the molecule AB at equilibrium, respectively and the second terms in Eqs. (47a) and (47b) refer to electron densities of each atom in the isolated systems A and B, respectively. The indices p and q are the number of atoms of the systems A and B, respectively.

5 Conclusions

In this article, we have reviewed the development of global and local reactivity descriptors which formed the backbone of the modern conceptual density functional theory. A principle global descriptors of chemical hardness has been used as a parameter for stability and this review has brought out the conditions under which the “principle of maximum hardness” operates. The global descriptors of hardness and softness have been related to molecule properties like polarisability. The review subsequently turns its attention to the local descriptors which can characterise the reactivity of a raison or a atom in a molecule. In this context, Fukui functions, local softness and their condensed versions, relative electrophilicity and nucleophilicity have been described. The review also highlights the importance of Hirshfeld analysis in obtaining non-negative condensed Fukui function as well as the more recent work on the effect of perturbation on these descriptors. Stability performance of these descriptors has also been discussed. These developments allow description of reactivity and molecule recommendation in terms of simple quantities and thus a potential use in chemical and biological interactions.

Acknowledgement The author acknowledges Mr. Susanta Das and Ms. Deepti Mishra for their help in preparation and critical proof reading of this manuscript. The author also acknowledges the help of Ms. Asha Shinde in providing secretarial assistance.

References

1. Pauling L (1967) The nature of chemical bond and structure of molecule and crystals. the nature of chemical bond and structure of molecule and crystals. Oxford New Delhi and IBH
2. Mc Weeny R (1979) Coulson's valence. Oxford University Press, Oxford
3. Maksic ZB (ed) (1990) Theoretical models of chemical bonding: the concept of the chemical bond. Springer, Berlin
4. Fukui K (1975) Theory of orientation and stereo selection. Springer, Berlin
5. Chermette H (1999) Chemical reactivity indexes in density functional theory. J Comput Chem 20(1):129–154
6. Parr RG, Yang W (1989) Density functional theory of atoms and molecules. Oxford University Press, New York
7. Chandra AK, Geerlings P, Nguyen MT (1997) On the asynchronism of isocyanide addition to dipolarophiles: application of local softness. J Org Chem 62(18):6417–6419
8. De Sekhar H, Krishnamurty S, Pal S (2010) Understanding the reactivity properties of Au_n $(6 < n < 13)$ clusters using density functional theory based reactivity descriptors. J Phys Chem C 114(14):6690–6703
9. Mignon P, Loverix S, Steyaert J, Geerlings P (2005) Influence of the π–π interaction on the hydrogen bonding capacity of stacked DNA/RNA bases. Nucleic Acids Res 33(6):1779–1789
10. Parthasarathi R, Subramanian V, Roy DR, Chattaraj PK (2004) Electrophilicity index as a possible descriptor of biological activity. Bioorg Med Chem 12(21):5533–5543
11. Chandra AK, Nguyen MT (2007) Use of DFT-based reactivity descriptors for rationalizing radical addition reactions: applicability and difficulties. Faraday Discuss 135:191–201
12. Geerlings P, Vivas-Reyes R, De PF, Biesemans M, Willem R (2003) DFT based reactivity descriptors and their application to the study of organotin compounds. NATO Sci Ser, II 116:461–495
13. Roy RK, Saha S (2010) Studies of regioselectivity of large molecular systems using DFT based reactivity descriptors. Annu Rep Prog Chem, Sect C: Phys Chem 106:118–162
14. Roy RK (2004) On the reliability of global and local electrophilicity descriptors. J Phys Chem A 108(22):4934–4939
15. Boon G, De PF, Langenaeker W, Geerlings P (1998) The use of density functional theory-based reactivity descriptors in molecular similarity calculations. Chem Phys Lett 295 (1,2):122–128
16. Pearson RG (1987) Recent advances in the concept of hard and soft acids and bases. J Chem Edu 64(7):561
17. Parr RG, Pearson RG (1983) Absolute hardness: companion parameter to absolute electronegativity. J Am Chem Soc 105(26):7512–7516
18. Yang W, Parr RG (1985) Proc Natl Acad Sci USA 82:6273
19. Pearson RG (1993) The principle of maximum hardness. Acc Chem Res 26(5):250
20. Harbola MK (1992) Magic numbers for metallic clusters and the principle of maximum hardness. Proc Natl Acad Sci USA 89(3):1036–1039
21. Pearson RG (1963) Hard and soft acids and bases. J Am Chem Soc 85(22):3533–3539
22. Sen KD (ed) (1993) Chemical hardness (structure and bonding), vol 80. Springer, Berlin
23. Chattaraj PK (ed) (2009) Ayers PW, Yang W, Bartolotti LJ. In chemical reactivity theory: A density functional view. Chemical reactivity theory. CRC; Boca Raton, Florida, pp. 255–268
24. Nalewajski RF (2005) Fukui function as correlation hole. Chem Phys Lett 410(4–6):335–338
25. Gal T, Geerlings P, De PF, Torrent-Sucarrat M (2011) A new approach to local hardness. Phys Chem Chem Phys 13(33):15003–15015
26. Chandra AK, Nguyen MT (2009) Fukui function and local softness as reactivity descriptors. In: Chattaraj PK (ed) Chemical reactivity theory: a density functional view. CRC, Boca Raton, FL, pp 163–178

27. Chandrakumar SKR, Pal S (2002) DFT and local reactivity descriptor studies on the nitrogen sorption selectivity from air by sodium and calcium exchanged zeolite-A Colloids and Surfaces, A 205 (1–2):127–138
28. Parr RG, Chattaraj PK (1991) Principle of maximum hardness. J Am Chem Soc 113 (5):1854–1855
29. Pal S, Vaval N, Roy R (1993) Principle of maximum hardness: an accurate ab initio study. J Phys Chem 97(17):4404–4406
30. Makov G (1995) Chemical hardness in density functional theory. J Phys Chem 99 (23):9337–9339
31. Sebastian KL (1994) On the proof of the principle of maximum hardness. Chem Phys Lett 231(1):40–42
32. Liu S, Parr RG (1997) Second-order density-functional description of molecules and chemical changes. J Chem Phys 106(13):5578–5586
33. Roy RK, Chandra AK, Pal S (1995) Hardness as a function of polarizability in a reaction profile. J Mol Struct (THEOCHEM) 331(3):261–265
34. Toro-Labbe A (1999) Characterization of chemical reactions from the profiles of energy, chemical potential, and hardness. J Phys Chem A 103(22):4398–4403
35. Roy R, Chandra AK, Pal S (1994) Correlation of polarizability, hardness, and electronegativity: polyatomic molecules. J Phys Chem 98(41):10447–10450
36. Gazquez JL (1997) Activation energies and softness additivity. J Phys Chem A 101 (48):8967–8969
37. Gazquez JL (1997) Bond energies and hardness differences. J Phys Chem A 101 (49):9464–9469
38. Pal S, Roy R, Chandra AK (1994) Change of hardness and chemical potential in chemical binding: a quantitative model. J Phys Chem 98(9):2314–2317
39. Kar R, Chandrakumar KRS, Pal S (2007) The Influence of electric field on the global and local reactivity descriptors: reactivity and stability of weakly bonded complexes. J Phys Chem A 111(2):375–383
40. Pal S, Chandra AK (1995) Some novel relationships of polarizability with dipole moments. J Phys Chem 99(38):13865–13867
41. Parr RG, Yang W (1984) Density functional approach to the frontier-electron theory of chemical reactivity. J Am Chem Soc 106:4049–4050
42. Yang W, Mortier WJ (1986) The use of global and local molecular parameters for the analysis of the gas-phase basicity of amines. J Am Chem Soc 108:5708–5711
43. Gazquez JL, Mendez F (1994) The hard and soft acids and bases principle: an atoms in molecules viewpoint. J Phys Chem 98(17):4591–4593
44. Mendez F, Gazquez JL (1994) Chemical reactivity of enolate ions: the local hard and soft acids and bases principle viewpoint. J Am Chem Soc 116(20):9298–9301
45. Langenaeker W, Coussement N, De Proft F, Geerlings P (1994) Quantum chemical study of the influence of isomorphous substitution on the catalytic activity of zeolites: an evaluation of reactivity indexes. J Phys Chem 98(11):3010–3014
46. Damoun S, Van de Woude G, Mendez F, Geerlings P (1997) Local softness as a regioselectivity indicator in [4 + 2] cycloaddition reactions. J Phys Chem A 101(5):886–893
47. Krishanmurthy S, Roy RK, Vetrivel R, Iwata S, Pal SJ (1997) The local hard – soft acid – base principle: a critical study. Phys Chem A 101:7253–7257
48. Chandra AK, Michalak A, Nguyen MT, Nalewajski RFJ (1998) Regional matching of atomic softnesses in chemical reactions: a two-reactant charge sensitivity study. Phys Chem A 102 (49):10182–10188
49. Chandra AK, Nguyen MT (1998) Density Functional Approach to Regiochemistry, Activation Energy, and Hardness Profile in 1,3-Dipolar Cycloadditions. J Phys Chem A 102 (30):6181–6185
50. Chandra AK, Nguyen MT (1997) J Chem Soc Perkin Trans 2:1415

51. Pal S, Chandrakumar KRS (2000) Critical study of local reactivity descriptors for weak interactions: an qualitative and quantitative analysis of adsorption of molecules in the Zeolite lattice. J Am Chem Soc 122(17):4145–4153
52. Chandrakumar KRS, Pal S (2002) Study of local hard-soft acid-base principle to multiple-site interactions. J Phys Chem A 106(23):5737–5744
53. Chandrakumar KRS, Pal S (2001) A novel theoretical model for molecular recognition of multiple-site interacting systems using density response functions. J Phys Chem B 105 (20):4541–4544
54. Roy RK, Pal S, Hirao K (1999) On non-negativity of Fukui function indices. J Chem Phys 110(17):8236–8245
55. Roy RK, Hirao K, Pal S (2000) On non-negativity of Fukui function indices. II. J Chem Phys 113(4):1372–1379
56. Tanwar A, Pal S (2005) Separability of local reactivity descriptors. J Chem Sci 117 (5):497–505
57. Roy RK, Krishnamurti S, Geerlings P, Pal S (1998) Local softness and hardness based reactivity descriptors for predicting intra- and intermolecular reactivity sequences: carbonyl compounds. J Phys Chem A 102(21):3746–3755
58. Tanwar A, Bagchi B, Pal S (2006) Interaction induced shifts in O–H stretching frequency of water in halide-ion water clusters: A microscopic approach with a bond descriptor. J Chem Phys 125(21):214304–214310
59. Hohenberg P, Kohn W (1964) Inhomogeneous electron gas. Phys Rev 136(3B):864–871
60. Sen KD, JØrgensen CK (eds) (1987) Electronegativity, Structure and Bonding, vol 66. Springer, Berlin
61. Mulliken RS (1934) New electroaffinity scale; together with data on valence states and on valence ionization potentials and electron affinities. J Chem Phys 2(11):782–794
62. Parr RG, Donnelly RA, Levy M, Palke WE (1978) Electronegativity: the density functional viewpoint. J Chem Phys 68(8):3801–3807
63. Pearson RG (1999) Maximum chemical and physical hardness. J Chem Educ 76(2):267–275
64. Pearson RG (1986) Proc Natl Acad Sci USA 83:8440
65. Huheey JE (1965) The electronegativity of groups. J Phys Chem 69(10):3284–3291
66. Sanderson RT (1952) An interpretation of bond lengths in alkali halide gas molecules. J Am Chem Soc 74:272–274
67. Sanderson RT (1976) Chemical bonds and bond energies. Academic, New York
68. Sanderson RT (1951) An interpretation of bond lengths and a classification of bonds. Science 114:670–672
69. Bultinck P, Langenaeker W, Lahorte P, De Proft F, Geerlings P, Waroquier M, Tollenaere JP (2002) The electronegativity equalization method i: parametrization and validation for atomic charge calculations. J Phys Chem A 106(34):7887–7894
70. Bultinck P, Langenaeker W, Lahorte P, De Proft F, Geerlings P, Van Alsenoy C, Tollenaere JP (2002) The electronegativity equalization method ii: applicability of different atomic charge schemes. J Phys Chem A 106(34):7895–7901
71. Chattaraj PK, Lee H, Parr RG (1991) HSAB principle. J Am Chem Soc 113(5):1855–1856
72. Datta D (1992) Hardness profile of a reaction path. J Phys Chem 96(6):2409–2410
73. Chattaraj PK, Liu GH, Parr RG (1995) The maximum hardness principle in the Gyftopoulos-Hatsopoulos three-level model for an atomic or molecular species and its positive and negative ions. Chem Phys Lett 237(1,2):171–176
74. Chattaraj PK, Nath S, Sannigrahi AB (1994) Hardness, chemical potential, and valency profiles of molecules under internal rotations. J Phys Chem 98(37):9143–9148
75. Cardenas-Jiron GI, Lahsen J, Toro-Labbe A (1995) Hardness profile and activation hardness for rotational isomerization processes. 1. application to nitrous acid and hydrogen persulfide. J Phys Chem 99(15):5325–5330

76. Cardenas-Jiron GI, Lahsen J, Toro-Labbe A (1995) Hardness profile and activation hardness for rotational isomerization processes. 2. The maximum hardness principle. J Phys Chem 99 (34):12730–12738
77. Gazquez JL, Martinez A, Mendez F (1993) Relationship between energy and hardness differences. J Phys Chem 97(16):4059–4063
78. Ghanty TK, Ghosh SK (1996) A density functional approach to hardness, polarizability, and valency of molecules in chemical reactions. J Phys Chem 100(30):12295–12298
79. Ghanty TK, Ghosh SK (2000) Molecular hardness, polarizability and valency variation of formamide and thioformamide on internal rotation: a density functional study. J Phys Chem A 104(13):2975–2979
80. Pearson RG, Palke WE (1992) Support for a principle of maximum hardness. J Phys Chem 96 (8):3283–3285
81. Chattaraj PK, Nath S, Sannigrahi AB (1993) Ab initio SCF study of maximum hardness and maximum molecular valency principles. Chem Phys Lett 212(3–4):223–230
82. Komorowski L (1987) Empirical evaluation of chemical hardness. Chem Phys Lett 134 (6):536–540
83. Komorowski L (1987) Electronegativity and hardness in the chemical approximation. Chem Phys 114(1):55–71
84. Sen KD, Bohm MC, Schmidt PC (1987) Electronegativity of atoms and molecular fragments. In: Sen KD (ed) Structure and Bonding, vol 66. Springer, Berlin, pp 99–123
85. Vinayagam SC, Sen KD (1988) CF_2^{2+} and CF^{2+}, two unusually stable dications with carbon-fluorine double bonding. Chem Phys Lett 144(2):178–181
86. Politzer P (1987) A relationship between the charge capacity and the hardness of neutral atoms and groups. J Chem Phys 86(2):1072–1073
87. Poltizer P, Grice MD, Murray JS (2001) Electronegativities, electrostatic potentials and covalent radii. J Mol Struct (THEOCHEM) 549(1–2):69–76
88. Politzer P, Huheey JE, Murray JS, Grodzicki M (1992) Electronegativities and electronaffinity. J Mol Struct (THEOCHEM) 259–265(4–5):99–104
89. Van Genechten K, Mortier WJ, Geerlings P (1987) Intrinsic framework electronegativity: a novel concept in solid state chemistry. J Chem Phys 86(9):5063–5071
90. Zunger A, Cohen ML (1979) First-principles nonlocal-pseudopotential approach in the density-functional formalism. II. Application to electronic and structural properties of solids. Phys Rev B 20(10):4082–4108
91. Poltizer P, Parr RG, Murphy DR (1983) Relationships between atomic chemical potentials, electrostatic potentials, and covalent radii. J Chem Phys 79(8):3859–3861
92. Ghanty TK, Ghosh SK (1996) New scale of atomic orbital radii and its relationship with polarizability, electronegativity, other atomic properties, and bond energies of diatomic molecules. J Phys Chem 100(30):17429–17433
93. Poltizer P, Parr RG, Murphy DR (1985) Approximate determination of Wigner-Seitz radii from free-atom wave functions. Phys Rev B 31(10):6809–6812
94. Ghanty TK, Ghosh SK (1994) Hardness and other properties. J Phys Chem 100 (21):8801–8807
95. Ghanty TK, Ghosh SK (1994) Simple density functional approach to polarizability, hardness, and covalent radius of atomic systems. J Phys Chem 98(37):9197–9201
96. Ghanty TK, Ghosh SK (1996) A new simple approach to the polarizability of atoms and ions using frontier orbitals from the Kohn-Sham density functional theory. J Mol Struct (THEOCHEM) 366(1–2):139–144
97. Ganguly P (1993) Simple interrelationship between crystal radii, pseudopotential orbital radii, and interatomic distances in elements. J Am Chem Soc 115(20):9287–9288
98. Perdew JP, Parr RG, Levy M, Balduz JL Jr (1982) Density-functional theory for fractional particle number: derivative discontinuities of the energy. Phys Rev Lett 49(23):1691–1694
99. Zhang Y, Yang W (2000) Perspective on "Density-functional theory for fractional particle number: derivative discontinuities of the energy". Theor Chim Acta 103(3):346–348

100. Nalewajski RF, Parr RG (2000) Information theory, atoms in molecules, and molecular similarity. Proc Natl Acad Sci 97(16):8879–8882
101. Ayers PW, Parr RG (2008) Local hardness equalization: Exploiting the ambiguity. J Chem Phys 128(18):184108–184116
102. Roy RK, Hirao K, Krishnamurty S, Pal S (2001) Mulliken population analysis based evaluation of condensed Fukui function indices using fractional molecular charge. J Chem Phys 115 (7):2901–2907
103. Mulliken RS (1955) Electronic population analysis on LCAO [Single Bond] MO molecular wave functions I. J Chem Phys 23(10):1833–1840
104. Breneman CM, Wiberg KB (1990) Determining atom-centered monopoles from molecular electrostatic potentials. The need for high sampling density in formamide conformational analysis. J Comput Chem 11(3):361–373
105. Reed AE, Curtiss LA, Weinhold F (1988) Intermolecular interactions from a natural bond orbital, donor-acceptor viewpoint. Chem Rev 88(6):899–926
106. Bonaccorsi R, Scrocco E, Tomasi J (1970) Molecular SCF calculations for the ground state of some three-membered ring molecules: $(CH_2)_3$, $(CH_2)_2NH$, $(CH_2)_2NH_2^+$, $(CH_2)_2O$, $(CH_2)_2S$, $(CH)_2CH_2$, and N_2CH_2. J Chem Phys 52(10):5270–5284
107. Cioslowski J (1989) A new population analysis based on atomic polar tensors. J Am Chem Soc 111(22):8333–8336
108. Cioslowski J, Martinov M, Mixon ST (1993) Atomic Fukui indexes from the topological theory of atoms in molecules applied to Hartree-Fock and correlated electron densities. J Phys Chem 97(42):10948–10951
109. Cioslowski J, Hay PJ, Ritchie JP (1990) Charge distributions and effective atomic charges in transition-metal complexes using generalized atomic polar tensors and topological analysis. J Phys Chem 94(1):148–151
110. Bader RFW (1990) Atoms in molecules: a quantum theory. Clarendon, Oxford
111. Bader RFW, Becker P (1988) Transferability of atomic properties and the theorem of Hohenberg and Kohn. Chem Phys Lett 148(5):452–458
112. Hirshfeld FL (1977) Bonded-atom fragments for describing molecular charge densities. Theor Chim Acta 44(2):129–138
113. Szabo A, Ostlund NS (1996) Modern quantum chemistry: introduction to advanced electronic structure theory. New York
114. Löwdin PO (1970) On the nonorthogonality problem. In: Löwdin PO (ed), Adv Quant Chem 5:185–199
115. Saha S, Roy RK, Ayers PW (2009) Are the Hirshfeld and Mulliken population analysis schemes consistent with chemical intuition? Int J Quantum Chem 109(9):1790–1806
116. John AP, Martin H-G, Krishnan R (1987) Quadratic configuration interaction. A general technique for determining electron correlation energies. J Chem Phys 87(10):5968–5975
117. Gustavo ES, Henry FS III (1989) Is coupled cluster singles and doubles (CCSD) more computationally intensive than quadratic configuration interaction (QCISD)? J Chem Phys 90(7):3700–3703
118. Arulmozhiraja S, Kolandaivel P (1997) Condensed Fukui function: dependency on atomic charges. Mol Phys 90(1):55–62
119. Misra GP, Sannigrahi AB (1996) A comparison of condensed Fukui function, free valency and unpaired spin population as reactivity indices for open-shell molecules. J Mol Struct (THEOCHEM) 361(1–3):63–68
120. Kar T, Sannigrahi AB (2000) Local reactivity indices of free radicals: Ab initio Hartree-Fock and Kohn-Sham density functional calculations. Ind J Chem A39:68–74
121. Nalewajski RF, Korchowiec J, Michalak A (1996) In density functional theory, IV. In: Nalewajski RF (ed) Topics in current chemistry, vol 183. Springer, Berlin, p 25
122. Parr RG, Parr JB (1999) Kenichi Fukui: recollections of a friendship. Theor Chim Acta 102 (1):4–6

123. Yang W, Parr RG, Pucci R (1984) Electron density, Kohn–Sham frontier orbitals, and Fukui functions. J Chem Phys 81(6):2862–2863
124. Flurchick K, Bartolotti L (1995) Visualizing properties of atomic and molecular systems. J Mol Graph 13(1):10–13
125. Chattaraj PK, Cedillo A, Parr RG (1995) Fukui function from a gradient expansion formula, and estimate of hardness and covalent radius for an atom. J Chem Phys 103(24):10621–10626
126. Pacios LF (1997) Study of a gradient expansion approach to compute the Fukui function in atoms. Chem Phys Lett 276(5–6):381–387
127. Pacios LF, Gómez PC (1998) Radial behavior of gradient expansion approximation to atomic Fukui function and shell structure of atoms. J Comput Chem 19(5):488–503
128. Chattaraj Pratim K, Andres C, Parr Robert G (1995) Variational method for determining the Fukui function and chemical hardness of an electronic system. J Chem Phys 103 (17):7645–7646
129. De Proft F, Geerlings P, Liu S, Parr RG (1999) Variational calculation of the global hardness and the Fukui Function via an approximation of the hardness Kernel. Pol J Chem 72:1737–1746
130. Michalak A, De Proft F, Geerlings P, Nalewajski RF (1999) Fukui functions from the relaxed Kohn Sham orbitals. J Phys Chem A 103(6):762–771
131. Roy RK, de Proft F, Geerlings P (1998) Site of protonation in aniline and substituted anilines in the gas phase: a study via the local hard and soft acids and bases concept. J Phys Chem A 102(35):7035–7040
132. Krishnamurty S, Pal S (2000) Intermolecular reactivity trends using the concept of group softness. J Phys Chem A 104(32):7639–7645
133. Shetty S, Kar R, Kanhere DG, Pal S (2005) Intercluster reactivity of metalloaromatic and antiaromatic compounds and their applications in molecular electronics: a theoretical investigation. J Phys Chem A 110(1):252–256
134. Pople JA (1976) Int J Quant Chem Symp 10:1
135. Pople JAE (1976) Structure, and reactivity. Wiley, New York
136. Cederbaum LS, Alon OE, Streltsov AI (2006) Coupled-cluster theory for systems of bosons in external traps. Phys Rev A 73(4):043609–043622
137. Geerlings P, De Proft F, Langenaeker W (2003) Conceptual density functional theory. Chem Rev 103(5):1793–1874
138. Fuentealba P, Cedillo A (1999) The variations of the hardness and the Kohn-Sham Fukui function under an external perturbation. J Chem Phys 110(20):9807–9811
139. Senet P (1996) Nonlinear electronic responses, Fukui functions and hardnesses as functionals of the ground-state electronic density. J Chem Phys 105(15):6471–6489
140. Senet P (1997) Kohn-Sham orbital formulation of the chemical electronic responses, including the hardness. J Chem Phys 107(7):2516–2524
141. Israelachvili J (1992) Intermolecular Surface Forces 2nd ed. London.
142. Bockris JOM, Reddy AKN (1973) Modern electrochemistry, vol 2. Plenum, New York
143. Sauer J (1989) Molecular models in ab initio studies of solids and surfaces: from ionic crystals and semiconductors to catalysts. Chem Rev 89(1):199–255
144. van Santen RA, Kramer GJ (1995) Reactivity theory of zeolitic broensted acidic sites. Chem Rev 95(3):637–660
145. Kreuzer HJ, Wang LC (1990) Field-induced surface chemistry of NO. J Chem Phys 93 (8):6065–6069
146. Ernst N, Drachsel W, Li Y, Block JH, Kreuzer HJ (1986) Field adsorption of helium on tungsten. Phys Rev Lett 57(21):2686–2689
147. Eckert M, Zundel G (1987) Proton polarizability, dipole moment, and proton transitions of an AH. . .B.dblharw. A-. . .H + B proton-transfer hydrogen bond as a function of an external electrical field: an ab initio SCF treatment. J Phys Chem 91(20):5170–5177
148. Hill TL (1958) Some possible biological effects of an electric field acting on nucleic acids or proteins. J Am Chem Soc 80(9):2142–2147

149. Kar R, Pal S (2008) Electric field response of molecular reactivity descriptors: a case study. Theor Chim Acta 120(4):375–383
150. Nalewajski RF (1984) Electrostatic effects in interactions between hard (soft) acids and bases. J Am Chem Soc 106(4):944–945
151. Li Y, Evans JNS (1995) The Fukui function: a key concept linking frontier molecular orbital theory and the hard-soft-acid-base principle. J Am Chem Soc 117(29):7756–7759
152. Gazquez JL (1997) The hard and soft acids and bases principle. J Phys Chem A 101 (26):4657–4659
153. Datta D (1992) On Pearson's HSAB principle. Inorg Chem 31(13):2797–2800
154. Dutta M (1999) The use of electronegativity and hardness towards understanding Pearson's hard-soft acid-base principle. Asian J Chem 11(1):198–202
155. Datta D, Singh SN (1991) Pearson's chemical hardness, heterolytic dissociative version of Pauling's bond-energy equation and a novel approach towards understanding Pearson's hard-soft acid-base principle. J Chem Soc Dalton Trans 6:1541–1549
156. Chattaraj PK, Schleyer PvR (1994) An ab initio study resulting in a greater understanding of the HSAB principle. J Am Chem Soc 116(3):1067–1071
157. Chattaraj PK, Gomez B, Chamorro E, Santos J, Fuentealba P (2001) Scrutiny of the HSAB principle in some representative acid base reactions. J Phys Chem A 105(38):8815–8820
158. Shoeib T, Gorelsky SI, Lever ABP, Siu KWM, Hopkinson AC (2001) When does the hard and soft acid base principle apply in the gas phase? Inorg Chim Acta 315(2):236–239
159. Shoeib T, El Aribi H, Siu KWM, Hopkinson AC (2001) A study of silver (I) ion organonitrile complexes: an ion structures, binding energies, and substituent effects. J Phys Chem A 105 (4):710–719
160. Chandrakumar KRS, Pal S (2003) Study of local hard-soft acid-base principle: effects of basis set, electron correlation, and the electron partitioning method. J Phys Chem A 107 (30):5755–5762
161. Chandrakumar KRS, Pal S (2002) The concept of density functional theory based descriptors and its relation with the reactivity of molecular systems. A semi-quantitative study. Int J Mol Sci 3(4):324–337
162. Mendez F, Gazquez JL (1994) The Fukui function of an atom in a molecule: a criterion for characterizing the reactive sites of chemical species. Proc Ind Acad Sci Chem Sci 106:183–193
163. Damoun S, Van de WG, Choho K, Geerlings P (1999) Influence of alkylating reagent softness on the regioselectivity in enolate ion alkylation: a theoretical local hard and soft acids and bases study. J Phys Chem A 103(39):7861–7866
164. Mendez F, Tamariz J, Geerlings P (1998) 1,3-dipolar cycloaddition reactions: A DFT and HSAB principle theoretical model. J Phys Chem A 102(31):6292–6296
165. Chandrakumar KRS, Pal S (2002) A systematic study on the reactivity of Lewis acid-base complexes through the local hard-soft acid-base principle. J Phys Chem A 106 (48):11775–11781

Struct Bond (2012) 149: 159–186
DOI: 10.1007/978-3-642-32753-7_5

Application of Reactivity Indices Within Density Functional Theory to Rationale Chemical Interactions

Abhijit Chatterjee

Abstract Chemistry is the science based on that all process involving bond making and bond breaking. Chemical interactions will determine the activity of the interacting species. If this reaction process can be mimicked by a handy and simple theory to test the validity this can be revolutionary. Reactivity index is that theory which was developed at the right time to rationalize chemical bonding. Density functional theory (DFT) has given precision to chemical concepts such as electronegativity, hardness, and softness and has embedded them in a perturbation approach to chemical reactivity. Since the majority of the reactions can be analyzed through the electrophilicity/nucleophilicity of various species involved, a proper understanding of these properties becomes essential. The hard soft acid–base (HSAB) principles classify the interaction between acids and bases in terms of global softness. In last few years the reactivity index methodology is well established and had found its application in a wide variety of systems. This study is to revisit the definition of reactivity index using DFT, within the domain of HSAB principle and then to discuss its application to rationale chemical interactions; in combination with intra- and intermolecular reactivity in materials.

Keywords Chemical bonding · Chemical interactions · DFT · Material · Reactivity index

Contents

A. Chatterjee (✉)
Accelrys K. K., Kasumigaseki Tokyu Bldg, 17F, 3-7-1 Kasumigaseki, Chiyoda-ku, Tokyo, Japan
e-mail: abhijit.chatterjee@accelrys.com

1 Introduction

Chemistry is the science based on that all process involving bond making and bond breaking. Chemical interactions will determine the activity of the interacting species. This had been long recognized that acids and bases theory remains the most important theory within the space of chemical bond and bonding. The history begins with the pioneering works of Lémery and Boyle in the early 1800 which introduced the "Solubility theory" and the associate principle of reactivity. Next, the major contribution of Rouelle came up, which prescribed the base concept as the complement of that of an acid, with his pneumatic theory of reactions, being culminated by the Lavoisier's contribution according to which the oxygen is directly related to the acidic character of matter. In 1900 Volta, Gay-Lussac and Liebig have preformed the historic physico-chemical experiments in elucidating the fact that acids have to contain hydrogen to be exchanged with a metal and a "radical" of different nature; they established the famous principle:

$$\text{acid} + \text{base} \leftrightarrow \text{salt} + \text{water}. \tag{1}$$

The first overall concept of acid–base was formulated by the Arrhenius, Van't Hoff, and Ostwald in the 1880s, leading to a picture where acids and bases release hydrogen and hydroxide ions, respectively, their interaction being responsible for the acid–base reactions. In the twentieth century, the acid and base definitions get a leap of the century when it happened to link with newly emerging quantum theory of atoms and molecules. The first theory belongs to Brønsted and Lowry [1] (1923), which assumes the proton as the particle, never free, which intermediates between an acid (the donor) and a base (the acceptor) compounds during chemical reactions. Within this framework the new acid–base interaction paradigm looks like:

$$\text{acid1} + \text{base2} \leftrightarrow \text{acid2} + \text{base1}. \tag{2}$$

Although efficient, this theory excessively enhanced the role of proton; fortunately, Lewis' intuition (1916) [2], the electron pair was soon recognized as a more

general conceptual tool in defining acids, bases, and their chemical bonding removing the unnecessary emphasis on protons. Worth noting, the Lewis base definition seems to superimpose on the Brønsted–Lowry theory while the acidic Lewis definition covers more general cases. The acid–base theory was refined again on the ground of the molecular orbital theory, derived by Pearson [3–8].

In this context, the chemical bonding and reactions were described in two steps: one step regards the Coulomb interaction, being quantified by the electronegativity index χ, seen as the negative of the chemical potential of the interacting systems [9], and by the associated equalization principle [10]; in the second step, the stability of the newly formed chemical bond is regulated by the so-called chemical hardness index η, seen as the second-order effect, consequently defined as the chemical force (i.e., the gradient of the chemical potential) acting on the bonding species [11–13]. In molecular orbital terms, the middle point of the HOMO–LUMO gap is associated with the chemical potential (i.e., minus electronegativity), while the weight of the gap is taken as the double of the chemical hardness of that molecule. With these, the acids and bases are further classified as soft ("*s*") and hard ("*h*"): a soft species has electrons easy to be transferred in the vacant orbital (LUMO) whereas the chemical reactions are more favorable as the HOMO level of one species vertically approaches the LUMO of the other. From now on, the molecular systems are recognized as hard and soft acids and bases (HSAB), in the sense that each molecule can be seen as hard–hard, soft–soft, hard–soft or soft–hard bonding combinations between acids and bases. The associate HSAB principle of chemical reactivity was formulated as well, providing that "hard acids prefer hard bases and soft acids prefer soft bases" [14, 15]:

$$h1 - s1 + s2 - h2 \leftrightarrow h1 - h2 + s2 - s1. \tag{3}$$

Despite the qualitative character [16–25] of the HSAB principle, an appropriate quantum index to smoothly distinguish between the soft and hard character of acids, bases, and their bonding, would switch HSAB toward a quantitative theory. A unified picture of the hard-and-soft-acids-and-bases and maximum hardness principles was approached through introducing the maximum hardness index Π. It provides particular chemical hardness ranges where the chemical bonding behaves like hard–hard, soft–soft, and hard–soft or soft–hard acid–base interaction characters and furnishes the key to analytical classification of acids and bases in an intrinsic structural manner. The reliability of the present recipe and index was tested by the chemical hardness ordering predictability and chemical bond nature characterization on a particular series of molecular Lewis acids and bases within various computational and experimental atomic chemical hardness scales. Although a consistent chemical hardness principles and related indices picture was furnished in all cases, considerable differences were noted with respect to the old-fashioned Pearson classification, which is one of the excellent work done by Putz et al. [26–34].

In the heterolytic cleavage of a bond, the electron pair lies with one of the fragments, which becomes electron-rich, while the other fragment becomes electron-deficient. An electron-rich reagent gets attracted to the center of the positive charge and forms a bond with an electron-deficient species by donating electrons. The electron-rich species is known as a nucleophile, and the electron-deficient one, as an electrophile [35–38]. Free radicals are generated through a corresponding homolytic process where an equal share of one electron is obtained by each fragment. Even radicals are designated as electrophilic/nucleophilic depending on their tendency to attack the reaction sites of relatively higher/lower electron density. Moreover, nucleophiles (electrophiles) are Lewis bases (acids) as well as reducing (oxidizing) agents since they donate (accept) electrons, implying a connection among electrophile–nucleophile chemistry, acid–base chemistry, and oxidation–reduction chemistry.

The chemical potential, chemical hardness and softness and reactivity indices have been used by a number of workers to assess a priori the reactivity of chemical species from their intrinsic electronic properties. The concept of electrophilicity has been known for several decades, although there has not been a rigorous definition of it until recently, Parr et al. [39] proposed a definition did they inspired by the experimental findings of Maynard et al. [40]. The revolution begins, with this simple index which has the ability to connect the major facets of chemical sciences.

Perhaps, one of the most successful and best-known methods is the frontier orbital theory of Fukui [41, 42]. Developed further by Parr and Yang [43], the method relates the reactivity of a molecule with respect to electrophilic or nucleophilic attack to the charge density arising from the highest occupied molecular orbital (HOMO) or lowest unoccupied molecular orbital (LUMO), respectively. According to the definition of global hardness, it is the second derivative of energy with respect to the number of electrons at constant temperature and external potential, which includes the nuclear field, whereas global softness is the inverse of global hardness. The hard soft acid–base (HSAB) principle [44], which was proposed by Pearson, classifies the interaction between acids and bases in terms of global softness. This HSAB principle can be applied successfully for various systems [6, 12, 17, 43, 45–54]. Furthermore, Pearson also suggested another principle of maximum hardness (PMH) [55]. It states that, for a constant external potential, the system with the maximum global hardness is most stable and also studied extensively to further probe into both inter- and intramolecular interactions [56–64]. Incorporation of HSAB concept into the DFT structure has several consequences, such as the hardness scale generated through the HSAB principle classifies chemical species in accordance with their behavior and in a good agreement with experimental values for a large number of cations, atoms, radicals and molecules [65]. On the other hand, Vela et al. suggested a linear relationship of global softness with dipole polarizability and consistent with empirical evidence [66]. In recent days, DFT has gained widespread interest in quantum chemistry. Some DFT-based local properties, e.g. Fukui functions and local softness as reactivity index have already been used for the reliable predictions in various types of electrophilic and nucleophilic reactive species involving in the chemical reactions [67, 68]. Moreover, the reactivity index scale can expand its domain successfully not only to predict the interaction between heteroatom and zeolite framework [69], but in various other fields [70–75].

This produces reliable results for the chemical properties of molecules and solids. DFT calculations are also fairly computationally inexpensive [75], making this the method of choice for accurate calculations on large molecular and solid-state systems. Finally, DFT programs for periodic systems are now widely available [76], making this the method of choice for modeling chemical reactivity on surfaces or within lattices. Such periodic calculations eliminate the uncertainties introduced by using finite-sized cluster approximations. In applications like the ones cited above, the magnitudes of the Fukui Functions (FFs) are correlated with the reactivity of various sites in a molecule. These FFs can be condensed to atomic-centered indices that can be used to predict which sites are most likely to react with electrophiles or nucleophiles. This can be used to compare the activities of sites within a molecule, or can be used as a measure of how various side groups alter the reactivity of a molecule.

The study of material properties based on the experiment is a difficult task because it is a complicated process. Theoretically, quantum mechanics is too simple to predict them. Hence, the application of reactivity index is an advanced and wise choice. Regarding material designing, prediction of the excited state has been considered a challenging and intricate problem as there is not much focus on chemical reactivity involving excited state [76]. In particular, the investigation of excited state properties are generally computed using time-dependent density functional theory (TD DFT). However, the disadvantages of TD DFT are notorious, which does not allow more accurate *ab initio* approaches than random phase approximation (RPA) and configuration interaction with single substitutions CIS. There is no privileged direction for improvement in DFT except changing the parameterized potential, which is not very promising alternative, considering the past experience with semi-empirical methods [77]. Moreover, the computational costs and the complexity experiences are comparable with RPA and CIS method. Hence, the calculation of excited state remains a challenging problem in DFT without involving perturbation. This provoked us to (1) revisit the fact that ground state DFT can reproduce the singlet excited state and (2) verify the process, until which extent the reactivity results can be reliable to explain the excited state behavior by comparing with experimental results. The reactivity index calculation of the ground state and excited state has been performed on closed systems such as methane, benzene and their chlorine substituted compounds. A comparison was made between the geometry of ground state and the excited state for those moieties through configuration interaction (CI) method with Austin Model 1 (AM1) Hamiltonian over the optimized geometry of DFT at the ground state. Results obtained through these two methodologies suggested that in terms of polarizability and heat of formation, DFT can reproduces the trend of excited state qualitatively. Again, those results can be further validated through UV spectral numbers, generated using CI method. The reactivity index proposition based on ground state was comparable with the excited state calculations and has potential to simulate the available experimental numbers.

With this background in this review we will now talk about the basic theory of the reactivity index including definitions of local and global softness along with relative nucleophilicity and electrophilicity. We will as well cover the role of response function in deriving the excited state reactivity index theory, followed by specific examples on chemical interactions with emphasis on cases where molecular interactions are detrimental for a chemical process.

2 Theory

In density functional theory, hardness (η) is defined as [13]

$$\eta = \frac{1}{2}\left(\frac{\delta^2 E}{\delta N^2}\right)v(r) = \frac{1}{2}\left(\frac{\delta\mu}{\delta N}\right)v, \tag{4}$$

where E is the total energy, N is the number of electrons of the chemical species and the chemical potential.

The global softness, S, is defined as the inverse of the global hardness (η):

$$S = \frac{1}{2\eta} = \left(\frac{\delta N}{\delta\mu}\right)v. \tag{5}$$

Using the finite difference approximation, S can be approximated as

$$S = \frac{1}{(\mathrm{IE} - \mathrm{EA})}, \tag{6}$$

where IE and EA are the first ionization energy and electron affinity of the molecule, respectively.

The Fukui function $f(r)$ is defined by [14]:

$$f(r) = \left[\frac{\delta\mu}{\delta v}(r)\right]_N = \left[\frac{\delta\rho(r)}{\delta N}\right]_v. \tag{7}$$

The function "f" is thus a local quantity, which has different values at different points in the species, N is the total number of electrons, μ is the chemical potential and v is the potential acting on an electron due to all nuclei present. Since $\rho(r)$ as a function of N has slope discontinuities, Eq. (4) provides the following three reaction indices [14]:

$$f^-(r) = \left[\frac{\delta\rho(r)}{\delta N}\right]v^- \quad \text{(governing electrophilic attack)},$$

$$f^+(r) = \left[\frac{\delta\rho(r)}{\delta N}\right]v^+ \quad \text{(governing nucleophilic attack)},$$

$$f^0(r) = \frac{1}{2}[f^+(r) + f^-(r)] \quad \text{(for radial attack)}.$$

In a finite difference approximation, the condensed Fukui function [14] of an atom, say x, in a molecule with N electrons is defined as:

$$fx^{+} = [q_x(N+1) - q_x(N)] \quad \text{(for nucleophilic attack)}, \tag{8}$$

$$fx^{-} = [q_x(N) - q_x(N-1)] \quad \text{(for electrophilic attack)},$$
$$fx^{0} = \frac{[q_x(N+1) - q_x(N)]}{2} \quad \text{(for radical attack)},$$

where q_x is the electronic population of atom x in a molecule.

The local softness $s(r)$ can be defined as

$$s(r) = \left(\frac{\delta\rho(r)}{\delta\mu}\right)_v. \tag{9}$$

Equation (6) can also be written as

$$s(r) = \left[\frac{\delta\rho(r)}{\delta N}\right] v \left[\frac{\delta N}{\delta\mu}\right]_v = f(r)S. \tag{10}$$

Thus, local softness contains the same information as the Fukui function $f(r)$ plus additional information about the total molecular softness, which is related to the global reactivity with respect to a reaction partner, as stated in HSAB principle. Thus the Fukui function may be therefore is thought of as a normalized local softness. Atomic softness values can easily be calculated by using Eq. (7), namely:

$$s_x{}^{+} = [q_x(N+1) - q_x(N)]S, \tag{11}$$

$$s_x{}^{-} = [q_x(N) - q_x(N-1)]S,$$

$$s_x^0 = \frac{S[q_x(N+1) - q_x(N-1)]}{2}.$$

In a recent work, Fitzgerald et al. [78] have shown that fractional charges as opposed to continuum charges can reduce the error in Fukui Index values up to 5 %. DFT is well-suited for use with noninteger occupations. Fractional occupations of orbital are commonly employed in the use of charge smearing to improve self-consistent field (SCF) convergence [79, 80]. Using fractional occupations, the partial derivatives are approximated as

$$f^{-}(\widehat{r}) = \left(\frac{\partial\rho(\widehat{r})}{\partial N}\right)_v^{-} \cong \frac{1}{\Delta N}\Big(\rho_N + \Delta(\widehat{r}) - \rho(\widehat{r})\Big), \tag{12}$$

$$f^{+}(\widehat{r}) = \left(\frac{\partial\rho(\widehat{r})}{\partial N}\right)_v^{+} \cong \frac{1}{\Delta N}\Big(\rho_N + \Delta(\widehat{r}) - \rho(\widehat{r})\Big). \tag{13}$$

There are some anomalous cases in which a specific atom shows both high electrophilicity and nucleophilicity due to the limitation of various basis set-dependent charge calculation procedures, and hence it is more appropriate to rationalize this concept of relative electrophilicity/nucleophilicity. Relative nucleophilicity is the nucleophilicity of a site relative to its own electrophilicity, and vice versa for relative electrophilicity. The idea of relative nucleophilicity/electrophilicity was first proposed by Roy et al. [81] to predict intramolecular reactivity sequences of carbonyl compounds. We have used a similar ratio for the first time to find the best di-octahedral smectite for nitrogen heterocyclics adsorption in terms of intermolecular interaction [82] and as well for the adsorption property of *para* and *meta* substituted nitrobenzene [83].

As the nuclear charge increases for a same number of electrons, the system become harder and more polarizable. A many particle system is completely characterized by total number of electrons (N) and the chemical potential $v(r)$ while χ and η describe the response of the system when N changes at fixed $v(r)$; the linear density response function $R(r, r')$ depicts the same for the variation of $v(r)$ for constant N [76]:

$$R(r\,r') = \left[\frac{\rho\delta(r)}{\delta v(r')}\right]N. \tag{14}$$

This response function can be expressed as

$$R(r\,r') = \left[\frac{s(r)s(r')}{S}\right] - s(r\,r'), \tag{15}$$

where $s(r, r')$, $s(r)$ and S are the softness kernel, local softness, and global softness. The linear response of the chemical species is measured in terms of static electric dipole polarizability in presence of weak external electric field.

It is important to note that the Fukui function and the related quantities may not provide proper reactivity trends for hard–hard interactions. Hard–hard interactions are charge-controlled since they are ionic in nature, whereas soft–soft interactions are frontier-controlled because of their covalent nature. The charge-based descriptors would be better suited to tackle the hard–hard interactions. During an electrophile–nucleophile interaction process, when two reactants approach each other from a large distance, they feel only the effect of the global electrophilicity of each other and not its local counterpart. Moreover, the numerical values of any condensed-to-atom quantity and the resulting trends should be analyzed with caution as they are empirical in nature owing to their dependence on the density partitioning scheme used.

3 Calculation Methodology

In the present study, all calculations have been carried out with DFT [84, 85] using $DMol^3$ code of Accelrys. A gradient-corrected functional BLYP [43, 86] and DNP basis set [87] was used throughout the calculation. Basis set superposition error

(BSSE) has also been calculated for the current basis set in nonlocal density approximation (NLDA) and the theories for reactivity index calculations were mentioned elsewhere in details [88]. Single-point calculations of the cation and anion of each molecule at the optimized geometry of the neutral molecules were also performed to evaluate Fukui functions as well as global and local softness. The condensed Fukui function and atomic softness was evaluated using Eqs. (8) and (11), respectively and the gross atomic charges were evaluated by the technique of electrostatic potential (ESP) driven charges. Geometries were optimized using analytic gradients and an efficient algorithm, which used delocalized internal coordinates so that the change of energy and the change of the maximum force was 2×10^{-5} Ha, and 0.004 Ha/Å, respectively. The studied molecule with a fixed symmetry was subjected to the electric field of 0.02 Hartree/Bohr parallel to the planar direction of the molecule. A field of ~1 V/Å has been applied to the molecules to calculate the response function.

Local Fukui Functions and global Softness were computed by finite differences with $\Delta N = 0.01$, 0.1 and 1.0. For each system, the DFT energy was converged to self-consistency. Atomic point charges were computed as described below. The SCF calculation was repeated using charges of $\Delta N = 0.01$, 0.1 and 1.0, and the atomic point charges were again computed, and condensed FFs were evaluated. DMol3, the program employs partition functions to divide space into regions associated with an atomic centre. Atomic charges were computed by integrating the charge density over all grid points while applying an appropriate partition function [78]:

$$q_k^H = Z_k - \sum_i \rho(\widehat{ri}) \frac{\rho_k(\widehat{ri})}{\rho_T(\widehat{ri})}, \tag{16}$$

where Z_k is the nuclear charge and ρ is the charge density of the isolated atom k,

$$\rho_T(\breve{r}_i) = \sum_j \rho_j(\breve{r}_i). \tag{17}$$

The sum over i runs over all numerical integration points in the molecule and the sum over j includes all atoms. This yields the Hirschfield atomic charges. We designate this type of atomic charge as q_k^H.

4 Application Examples

The reactivity index theory has been developed for the cause of chemical bonding and hence applied to all branches of chemical activity. There are different forms of describing the reactivity index, where the idea is to find the donor/acceptor capability of an atom present in a molecule interacting with another molecule or the interaction is within itself. This is the main concept, now depending on the interaction that is taking place; one can look into local softness of the atom, which is approaching the other interacting species or the group of atoms together approaching the active site. It has also been mentioned that if one wishes, one can describe the interaction between atoms for an intermolecular interaction through the concept of an equilibrium using the idea of reactivity index. Hence the concept reactivity index tells you the activity of atom center and its capability to interact with other species in its localized/nonlocalized neighbor. In recent years, various applications of reactivity index theory and its detailed description were studied [89]. According to the literature, two main issues are dealt with chemical reactivity index: (1) the chemical reactivity theory approach and its application for resolving chemical concern of importance within the helm of DFT and (2) application of DFT on resolving structure–property relationship in catalysis, reactions, and small molecules. With that background, this is the time again to show that how this theory can be applicable to address issues in industry and our main concerned industries are chemical, pharmaceutical, drug, semiconductor, and also polymers where people wanted to design molecule or material for a specific inter- or intramolecular interaction. Following are the few examples where chemical reactivity index can efficiently apply to shade light on the chemical interactions and rationalize scientific issues.

4.1 Scaling the Activity of Fluorophore Molecules

Anthracenes bearing aliphatic or aromatic amino substituent, which behave as molecular sensors, have shown their potential to act as photon-induced electron transfer (PET) systems. In this PET, the fluorophore moieties are responsible for electron release during protonation and deprotonation. The principle of HSAB deals with both intra- and intermolecular electron migration. It is possible to calculate the localized properties in terms of Fukui functions in the realm of DFT and thus calculate and establish a numerical matchmaking procedure that will generate an a priori rule for choosing the fluorophore in terms of its activity. We calculated the localized properties for neutral, anionic, and cationic systems to trace the course of the efficiency. A qualitative scale is proposed in terms of the feasibility of intramolecular hydrogen bonding. To investigate the effect of the environment of the nitrogen atom on protonation going from mono- to diprotonated systems the partial density of states has been calculated and compared the activity sequence with reactivity indices. The results show that location of the nitrogen atom in an aromatic ring does not influence the PET, but for aliphatic chains it plays a role. Furthermore, the protonation/deprotonation scenario has been explained. The results show that the reactivity indices can be used as a suitable property for scaling the activity of fluorophore molecules for the PET process [73].

4.2 Adsorption of Ozone-Depleting Chlorofluorocarbons

Adsorption of ozone-depleting chlorofluorocarbons (CFC) over zeolite is of major global environmental concern. To investigate the nature of CFCs including fluoro, chlorofluoro, and hydrofluoro/chloro carbons (CF_4, CF_3Cl, CF_2Cl_2, $CFCl_3$, CHF_3, $CHCl_3$) adsorption first-principle calculation was performed on faujasite models [88, 89]. Experimentally it is observed that separation of halocarbons is possible using Na–Y, though the cause is unknown. Reactivity index within the realm of HSAB principle was used to monitor the activity of the interacting CFCs using DFT to propose a qualitative order. The importance of both H-bonding and cation–F–Cl interactions in determining the low-energy sorption sites were monitored and rationalized. The host–guest interactions show a distinctive difference between the adsorption phenomenon between H–Y and Na–Y and as well for Cl and F. It is observed that Cl has more favorable interaction with hydrogen of H–Y compared to Na–Y and for F, the situation is reversed. To validate this trend periodic optimization calculations were performed. The interaction energy as obtained matches well with the reactivity index order resulted from cluster calculations. This study is a combination of DFT and periodic calculation to rationalize the electronic phenomenon of the chemical interaction process.

4.3 Designing of Stable Clay Nano-Composite

Resorcinol forms a novel nano-composite in the interlayer of montmorillonite. This resorcinol oligomer is stable inside the clay matrixes even above the boiling point of the monomer. A periodic ab initio calculation was performed with hydrated and nonhydrated montmorillonite before and after intercalation of resorcinol [90]. For the most feasible dimer- and tetramer-shaped oligomer of resorcinol, the intramolecular and intermolecular hydrogen bonding feasibility has been tested using the DFT-BLYP approach and the DNP basis set in the gas phase and in the presence of aqueous solvent. After locating the active site through Fukui functions within the realm of the HSAB principle, the relative nucleophilicity of the active cation sites in their hydrated state has been calculated. A novel quantitative scale in terms of the relative nucleophilicity and electrophilicity of the interacting resorcinol oligomers before and after solvation is proposed. Besides that, a comparison with a hydration situation and also the strength of the hydrogen bridges have been evaluated using mainly the dimer and cyclic tetramer type oligomers of resorcinol. In terms of localized reactivity index, the same atomic center of the resorcinol molecule produces the maximum electrophilicity and nucleophilicity. The electrophilicity is increasing after hydration and favors the interaction with the clay lattice with higher nucleophilicity. The monomers of resorcinol therefore can combine in the presence of water. As the monomers combine to form dimers or higher oligomers, their activity toward interaction with the clay interlayer increases. Localized reactivity calculation thus can propose the path of the interaction and its feasibility.

Using periodic ab-initio calculations, the formation mechanisms were traced: (1) resorcinol molecules combine without any interaction with water or (2) resorcinol oligomerizes through water. Both the mechanisms are compared and the effect of water on the process is elucidated. The results show that resorcinol molecules combine after hydration only and hence they are stable at higher temperature. The fittings of the oligomers were also tested as well by periodic calculation to compare the stability of the oligomers inside the newly formed clay nanocomposite.

4.4 Effect of Dopants on Bronsted and Lewis Acid Site

The influence of both bivalent and trivalent metal substituent from a range of metal cation (Co, Mn, Mg, Fe, and Cr) on the acidic property (both Brönsted and Lewis) of metal-substituted aluminum phosphate MeAlPOs is monitored [91]. The influence of the environment of the acid site is studied both by localized cluster and periodic calculations to propose that the acidity of AlPOs can be predictable with accuracy so that AlPO material with desired acidity can be designed. A semiquantitative reactivity scale within the domain of HSAB principle is proposed in terms of the metal substitutions using DFT. It is observed that for the bivalent metal cations Lewis acidity linearly increases with ionic size, whereas the Brönsted acidity is solely dependent on the nearest oxygen environment. Intramolecular and intermolecular interactions show that once the active site of the interacting species is identified, the influence of the environment can be prescribed. Mg^{+2}-doped AlPO-34 exhibits highest Brönsted acidity, whereas Cr^{+3}-doped species shows lowest acidity. Fe^{+2}–Fe^{+3}-doped AlPO-34 shows highest Lewis acidity, whereas Mn^{+2}, Mg^{+2} shows lowest acidity.

The cluster calculations were formed on localized cluster generated from the AlPO-34 structure with the terminal Al or P. Two independent clusters of the formula (1) $M^{+2}AlP_2O_{12}H_9$ and (2) $M^{+3}AlP_2O_{12}H_8$ generated by replacing one P by a M^{2+} or M^{3+} to represent the bivalent and trivalent dopant incorporated clusters respectively as shown in Fig. 1. The proton is included at the bridging oxygen where the dopant is incorporated for electrical neutrality for bivalent substitution. The terminal Al or P was replaced by hydrogen at that distance to mimic the real situation. It is observed that for the bivalent dopants the local environment is a distorted tetrahedral. For all the cases the M–OH distance is the longest. The M–O distances are considerably longer than the Al–O distance when the AlPO material is undoped, showing that the dopants introduce a considerable amount of distortion in the system. There is a drastic change in the M–O–P and angle values, ~135^0 compared to the Al–O–P ~148^0, which show that the observed structural distortion is not local and can be propagated beyond the nearest neighbor to the undoped region, which is in sharp contrast to the earlier results of Saadoune et al. [92]. To correlate the activity of dopants, hence we performed localized reactivity index calculation for the bivalent dopants using $M^{+2}AlP_2O_{12}H_9$ cluster. The Fukui function and local softness for the hydroxyl proton is presented both in terms of nucleophilic and electrophilic activity. Relative electrophilicity (s_x^+/s_x^-) and relative nucleophilicity (s_x^-/s_x^+) can be defined as the electrophilicity of any site as compared to its own (nucleophilicity for the first term and vice versa). The results are shown in Table 1. The cluster chosen was shown in Fig. 1.

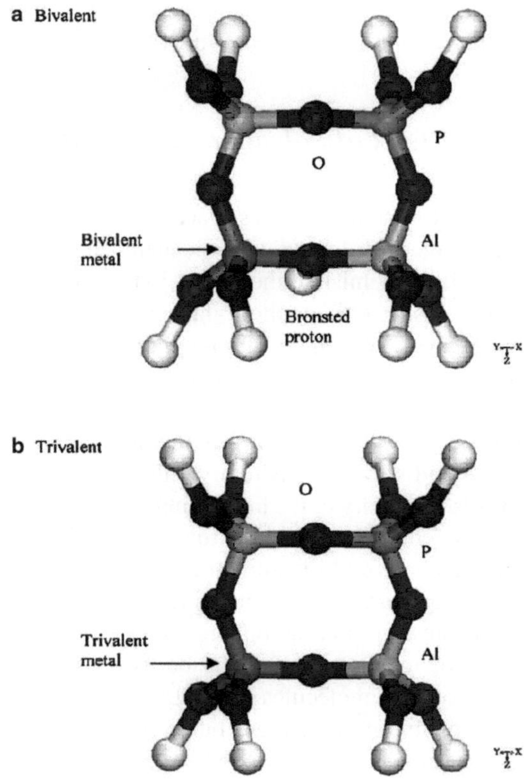

Fig. 1 Two independent cluster with the formula (**a**) $M^{+2}AlP_2O_{12}H_9$ and (**b**) $M^{+3}AlP_2O_{12}H_8$ to represent the bivalent and trivalent dopant incorporated clusters

Table 1 Local softness and relative nucleophilicity for the bivalent dopants calculated in terms of the hydroxyl proton using ESP charges by DFT to monitor Bronsted acidity trend

Metal ion	s_x^+	s_x^-	s_x^+/s_x^-
Mg^{2+}	0.20	0.53	2.555
Mn^{2+}	0.37	0.32	1.156
Cr^{2+}	0.43	0.44	0.977
Co^{2+}	0.53	0.36	1.478
Fe^{2+}	0.53	0.44	1.204

The results show that the relative nucleophilicity is highest for Mg^{2+} and is lowest for Cr^{2+}, which is opposite to the trend observed in terms of substitution energy. Fukui functions were used to monitor the dopant's activity in terms of Lewis acidity. For bivalent cation this order is totally different form the order obtained in terms of substitution energy and that obtained from the Bronsted acidity trend. For the trivalent dopant the highest and the lowest is for Mn^{3+}; the results match with the trend of substitution energy. The trend for Bronsted acidity is $Cr^{2+} < Mn^{2+} < Fe^{2+} < Co^{2+} < Mg^{2+}$, whereas the trend for Lewis acidity mainly for trivalent metal dopant is $Fe^{3+} > Co^{3+} > Cr^{3+} > Mn^{3+}$. This optimistic result encourages us to monitor a mixed valence situation, which may ideally exist during calcinations for the cations with variable oxidation state.

4.5 Effect of Solvation on the Interaction of Chromophore

Amino-functional silanol surface is mostly used for the immobilization of inorganic ions, molecules, organic, or biochemical molecules onto the mesopore surface. In analytical chemistry, the metal ion uptake was visualized through colorimetric sensors using chromophore molecules. One needs to know the structure–property correlation between the chromophore and silylating agent while choosing chromophore, which is very important to design the sensors. We have used two chromophores representative of hydrophobic and hydrophilic type and used density functional calculation on all the interacting molecules in both the unsolvated phase and solvated medium within the domain of HSAB principle to look at the localized activity of the interacting atoms of these reacting molecules to formulate a rule to choose the best chromophore. The mechanism of interaction between chromophore and the silylating agent has also been postulated. The results were compared with experiment and it is observed that solvation plays a detrimental role in the binding of chromophore with silylating agent. The results also show that the range of reactivity index can be used as a suitable property to scale activity of chromophore molecules suitable for the sensing process. It is observed that the hydrophobic chromophore binds stronger with both the metal and the silylating agent whereas for the hydrophilic one, it binds only with the silylating agent when solvated and in all cases the metal ion binding is weaker compared to that of the hydrophobic one [93].

In terms of global softness the order of activity for the chromophores both when solvated and unsolvated is as follows: PAR $>$ ZINCON. PAR shows much higher global softness than ZINCON, even compared to that of TMAC. In terms of dipole moment ZINCON shows greater hydrophilicity compared to TMAC when unsolvated. But TMAC is more hydrophilic than the ZINCON when solvated. At the same time it shows the greater hydrophobicity for PAR. We will now compare the atomic center of the silylating agent with highest relative electrophilicity/nucleophilicity to match with the counter active atom from chromophore with highest nucleophilicity/electrophilicity for pseudo bond formation. Based on these results one can foresee the interaction between the molecules as shown in Fig. 2 to propose the intramolecular chemical interaction.

4.6 Prediction of Interaction Between Metal Clusters with Oxide Surface

The HSAB principle classifies the interaction between acids and bases in terms of global softness. In last few years, the reactivity index methodology is well

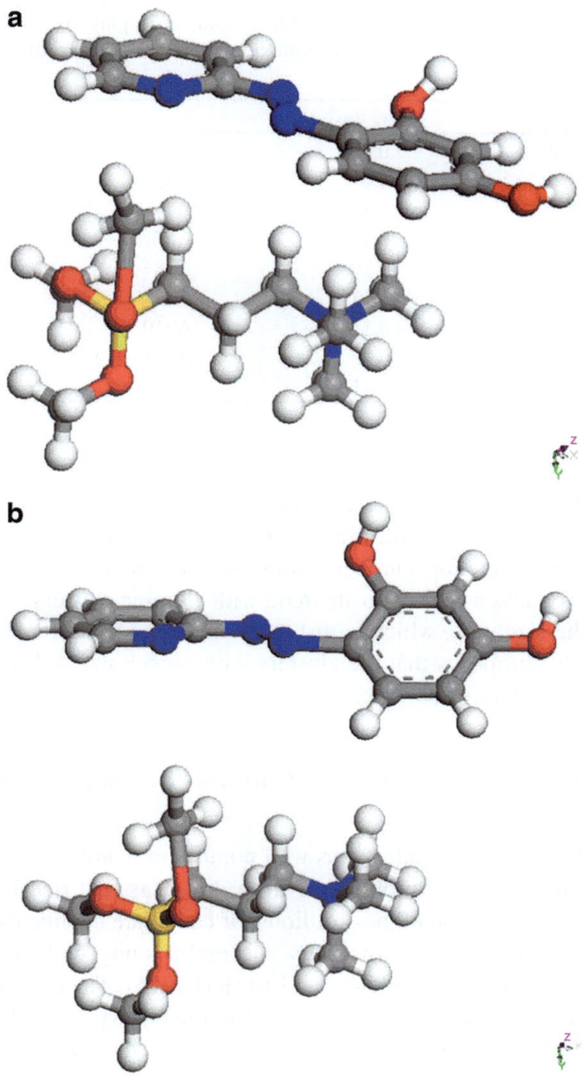

Fig. 2 Interaction between *N*-trimethoxysilylpropyl *N,N,N*-trimethyl ammonium chloride (TMAC) and 4-(2-pyridylazo)-resorcinol (PAR). (**a**) Unsolvated. (**b**) Solvated

established and had found its application in a wide variety of systems. This study deals with the viability of the reactivity index to monitor metal–cluster interaction with oxide. Pure gold cluster of a size between 2 and 12 was chosen to interact with clean alumina (100) surface. A scale was derived in terms of intra- and intermolecular interactions of gold cluster with alumina surface to rationalize the role of reactivity index in material designing [94].

We have calculated the relative nucleophilicity and electrophilicity for the clusters in the middle of the table, left hand side is for pure clusters and right hand side is for the clusters adsorbed over alumina surface.

Table 2 The relative electrophilicity and nucleophilicity of the individual atom centers in the gold cluster (left) before adsorption and the gold cluster after adsorption over alumina surface (right)

s_x^+/s_x^-	s_x^-/s_x^+	M	s_x^+/s_x^-	s_x^-/s_x^+
9.23	0.30	Au5(Td)	3.79	0.26
7.56	0.67	Au5(C4V)	3.73	0.27
6.91	0.14	Au6	4.43	0.22
5.38	0.18	Au7	4.11	0.24
5.30	0.19	Au8	5.36	0.19
5.76	0.17	Au9	6.86	0.15
5.66	0.15	Au10	8.71	0.11
5.37	0.16	Au11	11.51	0.09
5.17	0.11	Au12	15.00	0.07

From the results of Table 2 it is observed that the relative electrophilicity decreases with increase in the cluster size. That means that intramolecular interaction decreases with increase in cluster size as also justified by localized directive Fukui function numbers from the same table. The relative nucleophilicity really produces a very opposite trend with alumina clusters. The activity is increasing with the size now, which is just opposite to the trend for the pure gold cluster activity. This proposes that the clusters with bigger size will remain active after adsorbing over alumina surface.

4.7 Study on CDK2 Inhibitors Using Global Softness

The reactivity index is as well popular in pharmaceutical and drug applications. In particular, one problem of drug design is that one has to synthesize and screen thousands, sometimes millions, of candidate chemicals in developing one successful drug. There was a very successful study with reactivity index long back on human immunodeficiency virus (HIV) [95]. The cyclin-dependent kinases (CDKs) are a class of enzymes involved in the eukaryotic cell-cycle regulation.

A recent theoretical study on a series of CDK2 inhibitors used a set of global reactivity indices defined in terms of the density of states [96]. The related series were classified on the basis of the correlations obtained for the complete set of compounds and the sites targeted within the active site of CDK2. The comparison between the biological activity and the electronic chemical potential obtained through Fermi level yields poor results, thereby suggesting that the interaction between the hinge region of CDK2 and the ligands may have a marginal contribution from the charge transfer component. The comparison between the biological activity and global softness shows a better correlation, suggesting that polarization effects dominate over the CT contribution in the interaction between the so-called hinge region and the ligand. This result is very encouraging to show that the role of reactivity index in the intermolecular interaction, which can be further extrapolated to the intermolecular region to study the occupied states.

4.8 Molecular Interaction in Polymerization Reaction

Indeed, despite its many good characteristics, such as its long lifetime and the fact that this material is easy to process, the low thermal stability of PVC caused by the occurrence of side reactions in the polymerization process remains a problem. These side reactions lead to structural defects within the polymer, which have a great impact on the characteristics of the product. Better insight into the mechanism of these side reactions and their degree of reversibility would be helpful for improving the industrial production processes or, for PVC in particular, to reduce, for instance, the addition of environment-affecting thermal stabilizers during processing. A detailed investigation of the kinetic irreversibility–reversibility concept is presented on the basis of the analysis of four side reactions occurring in the polymerization of poly(vinyl chloride), the intramolecular 1,5- and 1,6-backbiting and 1,2- and 2,3-Cl shift side reactions. Density functional theory-based reactivity indices combined with an analysis of the reaction force are invoked to probe this concept. The reaction force analysis is used to partition the activation and reaction energy and characterize the behavior of reactivity indices along the three reaction regions that are defined within this approach. It has been observed that in the reactant and product regions mainly geometric rearrangements take place, whereas in the transition state region changes in the electronic bonding pattern occur; here most changes of the electronic properties are observed. The kinetic irreversibility–reversibility of the reactions is confirmed and linked to the differences in the Fukui function and dual descriptor of the radical centers associated with the initial and final species [97].

Allyl monomers are known as poor monomers to yield high molecular weight polymers via polymerization reactions [98]. The abstraction of the reactive allylic hydrogen of the monomer causes chain transfer reactions, which yield decreased molecular weight polymers [99]. Although allyl compounds are not good monomers for polymerization, their difunctional analogs can be polymerized through cyclopolymerization. In a recent work [100], various descriptors, defined within the framework of density functional theory (DFT), are used to explain the regioselectivity of the radical cyclizations preceding the intermolecular propagation step in the cyclopolymerization reactions. The transition states and the activation barriers for both the exo and the endo modes of the cyclization for a number of diallyl radicals are determined. An alternative and recently introduced energy decomposition of the activation barriers is used to investigate the steric effect in the cyclizations. Next, the non-spin-polarized and spin-polarized Fukui functions for a radical attack on the radical conformer minima close to the transition state are computed, in analogy with an earlier study of De Proft et al. [101, 102]. The reactive conformations of the radicals (designated as the reactive rotamers, not the structures corresponding to the global minima) are used for the calculation of the reactivity indices [103]. The regioselectivity in the cyclopolymerization of diallyl monomers is investigated using DFT-based reactivity indices. In the first part, the experimentally observed mode of cyclization (*exo* versus *endo*) of 11 selected radicals involved in this process is reproduced by the computation of activation energies, entropies, enthalpies, and Gibb's free energies for the five- and six-membered cyclization reactions. The application of a recently proposed energy partitioning of the activation barriers shows that the regioselectivity cannot be explained by the steric effect alone. Next, a number of relevant DFT-based reactivity indices, such as nonspin-polarized and spin-polarized Fukui functions, spin densities, and dual descriptors, were applied to probe the role of the polar and stereoelectronic effects in this reaction. The dual descriptor has been found to reproduce best the experimental trends, confirming the important role of the stereoelectronic effects.

The derivative of the Fukui function with respect to the number of electrons is the so-called dual descriptor of chemical reactivity, $f(r)$ [104],

$$f(r) = \left(\frac{\partial f(r)}{\partial N}\right) v = f^{+}(r) - f^{-}(r).$$

Among other things, the dual descriptor is useful for casting the famous Woodward–Hoffmann rules for pericyclic reactions in conceptual DFT [105]. $f(r)$ will be positive in regions of a molecule that are better at accepting electrons than they are at donating electrons, whereas $f(-)(r)$ will be negative in regions that are better at donating electrons than they are at accepting electrons. It is then stated that favorable chemical reactions occur when regions that are good electron acceptors ($f(r) > 0$) are aligned with regions that are good electron donors ($f(r) < 0$) [105].

4.9 Gas Sensor with Single-Wall Carbon Nanotube

In this part, we wish to explore interatomic interaction as well intramolecular interaction through the center of activity. Since the discovery of the structure of carbon nanotubes (CNTs) or single-walled nanotube (SWNT), much effort has been devoted to finding uses of these structures in applications ranging from filed-emission devices to other nano-devices [106, 107]. Kong et al. [108] proposed for the first time the use of CNTs as gas sensors. Experimental data have shown that transport properties of SWNT change dramatically upon exposure to gas molecules at ambient temperature [109]. Main advantage of the open SWNT bundles is that they provide a larger number of adsorption sites. As a result, the adsorption capacity is significantly increased and several new structures and phase transitions were observed [110]. A recent study of Andzelm et al. [111] indicate that the semiconducting SWNTs can serve as gas sensors for several gases like CO, NH_3, H_2, etc. However, NH_3 shows an intriguing behavior compared to other gasses. NH_3 molecule can bind weakly with CNTs, yet can change the conductance significantly. This discrepancy was explained by assuming that the NH_3 binds at defects. For a semiconducting SWNT exposed to 200 ppm of NO_2, it was found that the electrical conductance can increase by three orders of magnitude in a few seconds. On the other hand, exposure to 2 % NH_3 caused the conductance to decrease up to two orders of magnitude [112]. Sensors made from SWNT have high sensitivity and a fast response time at room temperature, which are important advantages for sensing applications. We have studied the interaction of CNT with different gas molecules such as O_2, N_2, H_2, CO_2, NO_2 to have an understanding of the adsorption behavior of the selected gases in defect-free CNT. We will as well focus on to figure out the effect of variation in the conductance with gas sorption by applying external electric field.

It is very difficult to obtain conductance by quantum mechanical calculation as it will be very much CPU intensive, but measurement of conductance is an utmost important parameter to prove the efficiency of the nanotubes as gas sensors, which is the experimental way of measuring sensors. Thus, a method was developed by calculating the change in the reactivity index before and after the application of the reaction field. The reactivity index provides information about the activity of the gas molecules over SWNT, if the activity changes then the sensing behavior will change [113]. This is a simplistic approach, which is cost-effective to new material design for the sensor industry.

4.10 Excited State Reactivity Index

This study aims to use the concept of ground-state reactivity index formalism within density functional theory (DFT) to predict the behavior of the excited state through the response function produced by weak electric field on chlorinated methanes and chlorinated benzenes. A comparison was made between the geometry of ground state and the excited state for those moieties through configuration interaction (CI) method with Austin Model 1 Hamiltonian over the optimized geometry of DFT at the ground state. Results obtained through these two methodologies suggested that in terms of polarizability and heat of formation, DFT can reproduce the excited state qualitatively. Again, those results can be further validated through UV spectral data, generated using CI method. The reactivity index proposition at ground state shows the potential of DFT to simulate excitation [114].

Primarily, to calculate the reactivity index at the ground state of both the closed systems; methane (CH_4) and benzene (C_6H_6) along with their chlorine-substituted compounds are chosen because methane and benzene is the doorway of the understanding of the properties of larger aliphatic and aromatic compound, respectively. A systematic substitution of hydrogen for both the moieties (methane and benzene) by chlorine was performed to get all the chlorine-substituted products until CCl_4 and C_6Cl_6, respectively. At first, those molecules are optimized with DFT at their neutral state and as well as cationic and anionic forms. The reactivity index values and the relative reactivity indices for individual centers of the series of chlorine-substituted methane and benzene-related compounds were computed. These molecules with the optimized structure were then subjected to the weak electric field to get the response function, followed by CI calculation with semi-empirical Hamiltonian. The mean polarizability, ionization potential, relative reactive indices, response function and heat of formation as obtained from DFT ground state calculation are compared to the CI excited state. The linear response of the electronic cloud of a chemical species to a weak external electric field is measured in terms of the static electric dipole polarizability. The obtained results were compared with the excited state results from regular CI method.

4.10.1 Absorption Spectra Calculations for Methane (CH_4) Series

All the geometry of CH_4 and chlorine-substituted CH_4 structures are optimized with DFT and then a PECI calculation has been performed. Consequently, the UV spectra of these series of molecules were obtained. The UV spectral data were compared with the experimental results.

4.10.2 Reactivity Index and Polarizability Calculation for Methane (CH_4) Series

The relative reactivity index is calculated, which is the electrophilicity of any site as compared to its own nucleophilicity for the first term and vice versa [74]; for the methane systems with varying amount of Chlorine replacing the hydrogen. The site with highest s_x^+/s_x^- ratio is the most probable site to be attacked by a nucleophile, and electrophilic attack is most feasible when s_x^-/s_x^+ ratio is highest. Those parameters can be used for both intermolecular and intramolecular interactions. To accomplish the change in the ground state activity of CH_4 and its chlorine-substituted compounds, the change of relative nucleophilicity and ionization potential (purely ground state property) was calculated and shown in Fig. 3a. A steady decrease in the relative nucleophilicity with the increasing chlorine substitution in the CH_4 moiety was observed, but there was no significant change in the ionization potential.

Then the polarizability through the response function was calculated using Eqs. (14) and (15) by applying the weak electric filed. The results are shown in Fig. 3b. Interestingly, a similar tendency in terms of polarizability as compared to that of relative nucleophilicity was also found. It has to be mentioned that in response function, a linear response of the electron cloud of a chemical species to the weak external field has been measured in terms of the static electric dipole polarizability, which was quite sensitive to the nature of the bonding, structure of the cluster and eventually related to the ionization potential. To justify the observation of polarizability behavior of a system resulted from the application of weak electric field, it was necessary to observe the polarizability changes in real excited state with multiple excitations.

4.10.3 Reactivity Index and Polarizability Calculation for Benzene (C_6H_6) Series

A steady decrease in the relative nucleophilicity with the increase of chlorine in the benzene moiety has been recognized. However, in contrast to the CH_4 series there

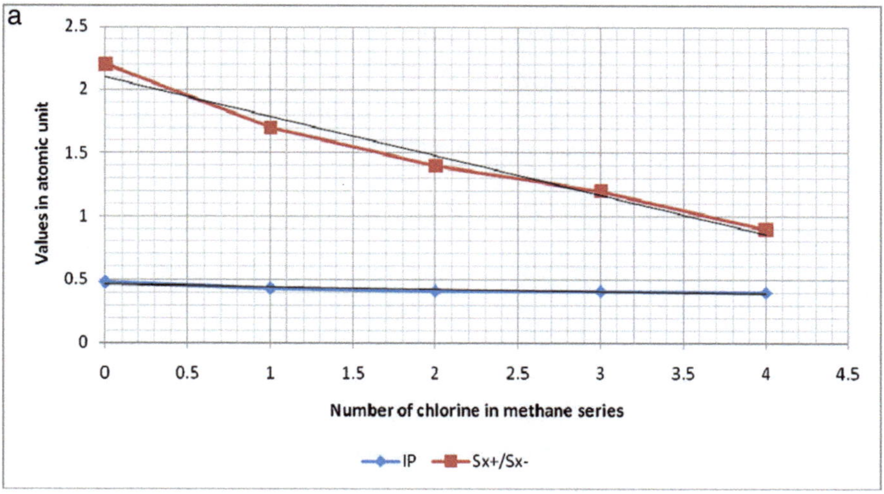

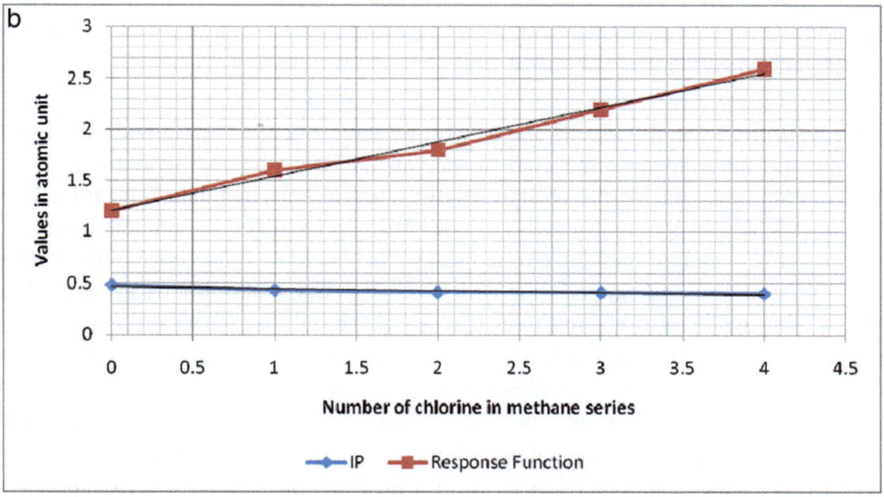

Fig. 3 (**a**) The relative nucleophilicity and ionization potential for the methane series with increase in the chlorine content. (**b**) The change in polarizability and ionization potential for the methane series with increase in the chlorine content, by applying electric field at the ground state

was no significant change in ionization potential with the increasing substitution of the chlorine observed. Moreover, the polarizability in terms of the response function gives a qualitative similarity as that of the relative nucleophilicity.

4.10.4 Comparison of Ground State and Excited State Reactivity Indices for Methane (CH_4) Series

In this communication we have so far performed ground state reactivity index calculation, excited state calculation with CI method whose credibility is validated

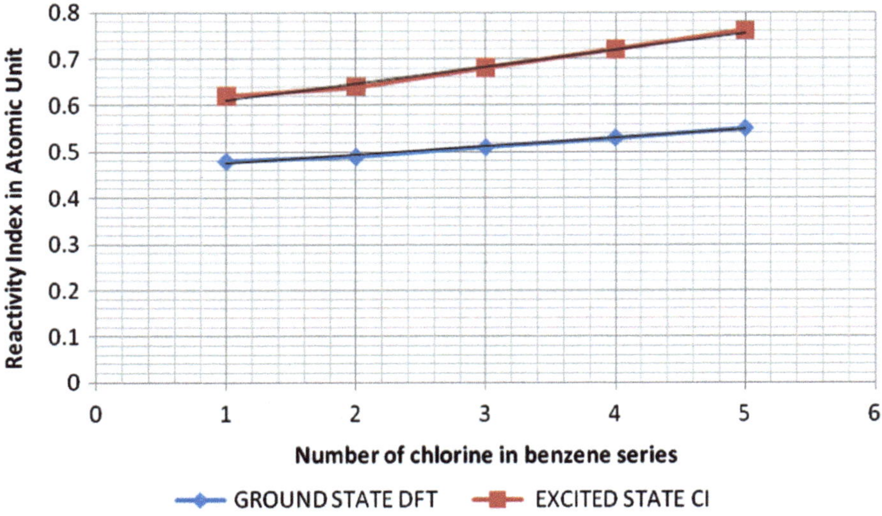

Fig. 4 The relative nucleophilicity as obtained for the benzene series with increase in the chlorine content for both ground state and excited state

with absorption spectral results and as well we have calculated polarizability using the electric field, keeping the goal to see can DFT will be able to reproduce the exited state behavior for the chosen small molecule here methane. To rationalize this, one would expect that reactivity index being a localized parameter should be compared between ground and excited state. The calculated reactivity indices both at the ground state and excited state with two different methodologies are plotted in Fig. 4. The behavioral change with respect to the carbon center present in CH_4 suggested an excellent match between DFT method (ground state) and CI method (excited state), even after the substitution of hydrogen by chlorine.

The test with methane and benzene analogous prove the hypothesis that DFT ground state can produce rationale numbers in excitation which can be comparable with excited state results by CI method. This methodology can now be extrapolated to a reasonable size system specifically for solar cell application with dye sensitized excitations with reasonable accuracy and in a faster calculation time with simple LCAO type DFT methodology.

5 Conclusion

In this review we have revisited the reactivity index theory from HSAB principle within the domain of DFT. We have presented an overview of the reactivity index theory from concept to industrial application. We have demonstrated that a theory within the DFT domain based on the theory of electronegativity and explored in the realm of electron affinity and ionization potential is capable to deliver a simple correlation to predict the intermolecular and intramolecular interaction. The theory is unique and we are able to describe the origin of the theory and its development down the years. If one can predict the localized interaction between interacting species carefully, then it will be possible to rationalize many chemical phenomena. We here have tried to share with you its capability through the various application examples from the research of our group and as well some recent applications to show that reactivity is an emerging area for material designing from nanocluster through nanowire, nanotube to biomaterial applications. The examples are from all different chemical interactions occurring between different molecular domains. We have shown that the robustness of the theory in terms of its extendibility form localized interaction, to global interactions and as well to a relative scale of implementation. We also have explored the recent development of this theory in excited state. One can therefore use this theory and the indices to rationalize chemical bonding and hence will have the capability to prescribe interactions detrimental for a success of chemical process. We conclude with an optimistic note that the electrophilicity/nucleophilicity will grow from its current tremendous predictive potential, in combination with the related property information for chemical bonding like charge density, orbital overlap integrals, and therefore will be adequate in developing a more expandable theory of chemical interaction removing its limit to reactivity.

References

1. Bronsted JN (1923) Some remarks on the concept of acids and bases, Recueil des Travaux Chimiques des Pays-Bas Volume 42:718–728
2. Lewis GN (1916) The atom and the molecule. J Am Chem Soc 38:762–785
3. Pearson RG (1963) Hard and soft acids and bases. J Am Chem Soc 85:3533–3539
4. Pearson RG (1985) Absolute electronegativity and absolute hardness of Lewis acids and bases. J Am Chem Soc 107:6801–6806
5. Pearson RG (1986) Absolute electronegativity and hardness correlated with molecular orbital Theory. Proc Natl Acad Sci USA 83:8440–8441
6. Pearson RG (1987) Recent advances in the concept of hard and soft acids and bases. J Chem Edu 64:561–567
7. Pearson RG (1988) Absolute electronegativity and hardness: Application to inorganic chemistry. Inorg Chem 27:734–740
8. Pearson RG (1989) Absolute electronegativity and hardness: Application to organic chemistry. Org Chem 54:1423–1430
9. Pearson RG (1997) Chemical Hardness. Wiley-VCH, Weinheim
10. Kohn W, Becke AD, Parr RG (1996) Density functional theory of electronic structure. J Phys Chem 100:12974–12980
11. Putz M.V., (2003) Contributions within Density Functional Theory with Applications in chemical reactivity theory and electronegativity, Dissertation. Com, Parkland, Florida
12. Parr RG, Pearson RG (1983) Absolute hardness: Companion parameter to absolute electronegativity. J Am Chem Soc 105:7512–7516
13. Chattaraj PK, Parr RG (1993) Density functional theory of chemical hardness. Struct Bond 80:11–25
14. Parr RG, Gázquez JL (1993) Hardness functional. J Phys Chem 97:3939–3940

15. Zhou Z, Parr RG (1990) Activation hardness: New index for describing the orientation of electrophilic aromatic substitution. J Am Chem Soc 112:5720–5724
16. Drago RS, Kabler RA (1972) A Quantitative evaluation of the HSAB concept. Inorg Chem 11:3144–3145
17. Chattaraj PK, Lee H, Parr RG (1991) HSAB principle. J Am Chem Soc 113:1855–1856
18. Nalewajski RF (1984) Electrostatic effects in interaction between hard (soft) acids and bases. J Am Chem Soc 106:944–945
19. Chattaraj PK, Sengupta S (1996) Popular electronic structure principles in a dynamical context. J Phys Chem 100:16129–16130
20. Komorowski L, Boyd SL, Boyd RJ (1996) Electronegativity and hardness of disjoint and transferable molecular fragments. J Phys Chem 100:3448–3453
21. Ponti A (2000) DFT-based regioselectivity criteria for cycloaddition reaction. J Phys Chem 104:8843–8846
22. Chandrakumar KRS, Pal S (2002) Study of local hard-soft acid–base principle to multiple-site interactions. J Phys Chem A 106:5737–5744
23. Chandrakumar KRS, Pal S (2002) A systematic study of the reactivity of Lewis acid–base complexes through the local hard-soft acid–base principle. J Phys Chem A 106:11775–11781
24. Chandrakumar KRS, Pal S (2002) Study of local hard-soft acid–base principle: Effects of basis set, electron correlation, and the electron partitioning method. J Phys Chem A 107:5755–5762
25. Chattaraj PK, Maiti B (2003) HSAB principle applied to the time evolution of chemical reactions. J Am Chem Soc 125:2705–2710
26. Putz MV, Russo N, Sicilia E (2004) On the application of the HSAB principle through the use of improved computational schemes for chemical hardness evaluation. J Comput Chem 25:994–1003
27. Torrent-Sucarrat M, Luis JM, Duran M, Solà M (2005) An assessment of a simple hardness kernel approximation for the calculation of the global hardness in a series of Lewis acids and bases. J Mol Struct (THEOCHEM) 727:139–148
28. Kleinert H, Pelster A, Putz MV (2002) Variational perturbation theory for Markov processes. Phys Rev E 65:066128
29. Putz MV, Russo N, Sicilia E (2003) Atomic radii scale and related size properties from density functional electronegativity formulation. J Phys Chem A 107:5461–5465
30. Putz MV (2005) Markovian approach of the electron localization functions. Int J Quantum Chem 105:1–11
31. Putz MV (2006) Systematic formulations for electronegativity and hardness and their atomic scales within density functional softness theory. Int J Quantum Chem 106:361–389
32. Putz MV (2007) Semiclassical electronegativity and chemical hardness. J Theor Comp Chem 6:33–47
33. Putz MV (2008) Maximum hardness index of quantum acid–base bonding. Math Commun Math Comput Chem 60:845–868
34. Putz MV (2009) Electronegativity: quantum observable. Int J Quantum Chem 109:733–738
35. Carey FA, Sundberg RJ (2001) Adv Org Chem; Part B: reactions and synthesis, 4th ed.; Kluwer Academic/Plenum Publishers: New York
36. Lowry TH, Richardson KS (1987) Mechanism and theory in organic chemistry, 3rd edn. Harper & Row, New York
37. Smith MB, March J (2001) Advanced organic chemistry: reactions, mechanisms, and structure, 5th edn. Wiley, New York
38. Sykes P (1970) A guidebook to mechanism in organic chemistry, 6th edn. Orient Longman Limited, New Delhi
39. Parr RG, Szentpaly LV, Liu S (1999) Electrophillicity index. J Am Chem Soc 121:1922–1924

40. Maynard AT, Huang M, Rice WG, Covell DG (1998) Reactivity of the HIV-1 nucleocapsid protein p7 zinc finger domains from the perspective of density-functional theory. Proc Natl Acad Sci U S A 95:11578–11583
41. Fukui K (1973) Theory of orientation and stereoselection. Springer, Berlin
42. Fukui K (1982) Role of frontier orbital in chemical reactions. Science 218:747–754
43. Parr RG, Yang W (1984) Density functional approach to the frontier-electron theory of chemical reactivity. J Am Chem Soc 106:4049–4050
44. Chattaraj PK, Poddar A (1999) Chemical reactivity and excited-state density functional theory. J Phys Chem A 103:1274–1275
45. Geerlings P, De Proft F (2000) HSAB principle: Applications of its global and local forms in organic chemistry. Int J Quantum Chem 80:227–235
46. Ayers PW, Levy M (2000) Perspective on density functional approach to the frontier-electron theory of chemical reactivity. Theor Chem Acc 103:353–360
47. Baeten A, Tafazoli M, Kirsch-Volders M, Geerlings P (1999) Use of the HSAB principle in quantitative structure-activity relationships in toxicological research: Application to the genotoxicity of chlorinated hydrocarbons. Int J Quantum Chem 74:351–355
48. Mendez FDL, Romero M, De Proft F, Geerlings P (1998) The basicity of p-substituted phenolates and the elimination-substitution ratio in p-nitrophenethyl bromide: A HSAB theoretical study. J Org Chem 63:5774–5778
49. Mendez F, Tamariz J, Geerlings P (1998) 1,3-dipolar cycloaddition reactions: A DFT and HSAB principle theoretical model J. Phys Chem A 102:6292–6296
50. Ayers PW, Parr RG, Pearson RG (2006) Elucidating the hard/soft acid/base principle: A perspective based on half-reactions. J Chem Phys 124:194107
51. Ayers PW (2005) An elementary derivation of the hard/soft acid/base principle. J Chem Phys 122:1–3
52. Ayers PW (2007) The physical basis of the global and local hard/soft acid/base principles. Faraday Discuss 135:161–190
53. Parr RG, Chattaraj PK (1991) Principle of maximum hardness. J Am Chem Soc 113:1854–1855
54. Chattaraj PK (1996) Proc Ind Natl Sci Acad Part A 62:513
55. Chattaraj PK, Parr RG (1993) Struct Bonding (Berlin) 80:11
56. Ayers PW, Parr RG (2000) Variational principles for describing chemical reactions: The Fukui function and chemical hardness revisited. J Am Chem Soc 122:2010–2018
57. Torrent-Sucarrat M, Luis JM, Duran M, Sola M (2002) Are the maximum hardness and minimum polarizability principles always obeyed in nontotally symmetric vibrations? J Chem Phys 117:10561–10570
58. Torrent-Sucarrat M, Luis JM, Duran M, Sola M (2001) On the validity of the maximum hardness and minimum polarizability principles for nontotally symmetric vibrations. J Am Chem Soc 123:7951–7952
59. Robles J, Bartolotti LJ (1984) Electronegativities, electron-affinities, ionization potentials and hardness of the elements within spin polarized density functional theory. J Am Chem Soc 106:3723–3727
60. Parr RG, Yang WT (1995) Density functional theory of the electronic-structure of molecules. Ann Rev Phys Chem 46:701–728
61. Geerlings P, De Proft F, Langenaeker W (2003) Conceptual density functional theory. Chem Rev 103:1793–1873
62. Pearson RG (1988) Absolute electronegativity and hardness: application to inorganic chemistry. Inorg Chem 27:734–737
63. Parr RG, Yang WT (1989) Density-functional theory of atoms and molecules. Oxford University Press, New York
64. Vela A, Gazquez JL (1990) A relationship between the static dipole polarizability, the global softness and the Fukui function. J Am Chem Soc 112:1490–1492

65. Chermette H (1999) Chemical reactivity indexes in density functional theory. J Comput Chem 20:129–154
66. Chattaraj PK, Giri S, Duley S (2011) Chem Rev 111:43–75
67. Ayers PW, Anderson JSM, Bartolotti LJ (2005) Perturbative perspectives on the chemical reaction prediction problem. Int J Quantum Chem 101:520–534
68. Chatterjee A, Iwasaki T, Ebina T (1999) Reactivity index scale for interaction of heteroatomic molecules with zeolite framework. J Phys Chem A 103:2489–2494
69. Chatterjee A, Iwasaki T, Ebina T (2001) Best dioctahedral smectite for nitrogen heterocyclics adsorption—A reactivity *index* study. J Phys Chem A 105:10694–10701
70. Chatterjee A, Iwasaki T (2001) A reactivity index study to choose the best template for a particular zeolite synthesis. J Phys Chem A 105:6187–6194
71. Chatterjee A, Iwasaki T, Ebina T, Mizukami F (2003) Intermolecular reactivity studyto scaleadsorptionproperty of para- and meta-substituted dnitrobenzene over 2: 1 dioctahedral smectite. J Chem Phys 118:10212–10220
72. Chatterjee A, Onodera Y, Ebina T, Mizukami F (2003) 2,3,7,8-tetrachloro dibenzo-p-dioxin can be successfully decomposed over 2: 1 dioctahedral smectite—a reactivity index study. J Mol Graph Model 22:93–104
73. Chatterjee A, Suzuki T, Onodera Y, Tanaka DAP (2003) A density functional study to choose the best fluorophore for photon-induced electron-transfer (PET) sensors. Chem Eur J 9:3920–3929
74. Serrano-Andres L, Merchan M (2005) Quantum chemistry of the excited state: 2005 overview. J Mol Struct Theochem 729:99–108
75. Fitzerald G (2008) Molecular simulation, 34, (10–15):931–936
76. Kresse G, Furthmuller J (1996) Efficiency of ab-initio total energy calculations for metals and semiconductors using a plane-wave basis set. Comput Math Sci 6:15–50
77. Rabuck AD, Scuseria GE (1999) Improving self-consistent field convergence by varying occupation numbers. J Chem Phys 110:695–700
78. Roy RK, Krishnamurti S, Geerlings P, Pal S (1998) Local softness and hardness based reactivity descriptors for predicting intra- and intermolecular reactivity sequences: Carbonyl compounds. J Phys Chem A 102:3746–3755
79. Delley B (1990) An all-electron numerical method for solving the local density functional for polyatomic molecules. J Chem Phys 92:508–517
80. Delley B (2000) From molecules to solids with the DMol3 approach. J Chem Phys 113:7756–7764
81. Delley B (1991) Analytical energy derivatives in the numerical local density functional approach. J Chem Phys 94:7245–7250
82. Becke AD (1988) A multicenter numerical-integration scheme for polyatomic molecules. J Chem Phys 88:2547–2553
83. Lee CT, Yang WT, ParrR G (1988) Development of the Colle-Salvetti correlation-energy formula into a functional of the electron density. Phys Rev B 37:785–789
84. Chatterjee A, Onodera Y, Ebina T, Mizukami F (2004) Effect of exchangeable cation on the swelling property of 2: 1 dioctahedral smectite—a periodic first principle study. J Chem Phys 120:3414–3424
85. Berkowitz M, Ghosh SK, Parr RG (1985) On the concept of local hardness in chemistry. J Am Chem Soc 107:6811–6814
86. Yang W, Mortier WJ (1986) The use of global and local molecular parameters for the analysis of the gas-phase basicity of amines. J Am Chem Soc 108:5708–5711
87. Chatterjee A (2002) Edited the special issue on application of density functional theory in chemical reactions. Int J Mol Sci 4:234–444
88. Chatterjee A, Ebina T, Iwasaki T (2003) Adsorption structures and energetic of fluoro- and chlorofluorocarbons over faujasite—a first principle study. Stud Surf Sci Catal 145:371–374

89. Chatterjee A, Ebina T, Iwasaki T, Mizukami F (2003) Cholorofluorocarbons adsorption structures and energetic over faujasite type zeolites—a first principle study. THEOCHEM 630:233–242
90. Chatterjee A, Ebina T, Mizukami F (2005) Effects of water on the structure and bonding of resorcinol in the interlayer of montmorillonite nanocomposite—a periodic first principle study. J Phys Chem B 109:7306–7313
91. Chatterjee A (2006) A reactivity index study to rationalize the effect of dopants on Brönsted and Lewis acidity occurring in MeAlPOs. J Mol Graph Model 24:262–270
92. Saadoune I, Cora F, Catlow CRA (2003) Computational study of the structural and electronic properties of dopant ions in microporous AlPOs. 1. Acid catalytic activity of divalent metal ions. J Phys Chem B 107:3003–3011
93. Chatterjee A, Balaji T, Matsunaga H, Mizukami F (2006) A reactivity index study to monitor the role of solvation on the interaction of the chromophores with amino-functional silanol surface for colorimetric sensors. J Mol Graph Model 25:208–218
94. Chatterjee A, Kawazoe A (2007) Application of the reactivity index to propose intra and intermolecular reactivity in metal cluster interaction over oxide surface. Mater Trans 48:2152–2158
95. Maynard AT, Huang M, Rice WG, Covell DG (1998) Reactivity of the HIV-1 nucleocapsid protein p7 zinc finger domains from the perspective of density-functional theory. Proc Natl Acad Sci USA 95:11578–11583
96. Renato C, Alzate-Morales JH, William T, Santos Juan C, Ca'rdenas C (2007) AQ4 Theoretical study on CDK2 inhibitors using a global softness obtained from the density of states. J Phys Chem B 111:3293–3297
97. Vleeschouwer FD, Toro-Labbe A, Gutie´rrez-Oliva S, Speybroeck V, Waroquier M, Geerlings P, De Proft F (2009) Reversibility from DFT-based reactivity indices: intramolecular side reactions in the polymerization of poly(vinyl chloride). J Phys Chem A 113:7899–7908
98. Zubov VP, Kumar MV, Masterova MN, Kabanov VA (1979) Reactivity of allyl monomers in radical polymerization. J Macromol Sci Chem 1:111–131
99. Matsumoto A (2001) Polymerization of multiallyl monomers. Prog Polym Sci 26:189–257
100. Harada S, Hasegawa S (1984) Homopolymerization of monoallyl ammonium salts with azo initiator. Macromol Chem Rapid Commun 5:27–31
101. Ugur I, Vleeschouwer FD, Tuzun N, Aviyente V, Geerlings P, Liu S, Ayers PW, Proft FD (2009) Cyclopolymerization reactions of diallyl monomers: exploring electronic and steric effects using DFT reactivity indices. J Phys Chem A 113:8704–8711
102. Pinter B, De Proft F, Van Speybroeck V, Hemelsoet K, Waroquier M, Chamorro E, Veszpremi T, Geerlings P (2007) Spin-polarized conceptual density functional theory study of the regioselectivity in ring closures of radicals. J Org Chem 72:348–356
103. Kodaira T, Liu Q-Q, Satoyama M, Urushisaki M, Utsumi H (1999) Cyclopolymerization - XXVI. Repeating unit structure of cyclopolymers derived from N-substituted-N-allyl-2-(methoxycarbonyl)allylamines and mechanism of intramolecular cyclization. Polymer 40:6947–6954, and references therein
104. Morell C, Grand A, Toro-Labbe A (2005) New dual descriptor for chemical reactivity. J Phys Chem A 109:205–212
105. Ayers PW, Morell C, De Proft F, Geerlings P (2007) Understanding the Woodward-Hoffmann rules by using changes in electron density. Chem Eur J 13:8240–8247
106. Ijima S (1991) Helical microtubules of graphitic carbon. Nature 354:56–58
107. De Heer WA, Chatelain A, Ugarte D (1995) A carbon nanotube field-emission electron source. Science 270:1179–1180
108. Kong J, Franklin NR, Zhou CW, Chapline MG, Peng S, Cho KJ, Dai HJ (2000) Nanotube molecular wires as chemical sensors. Science 287:622–625
109. Collins PG, Bradley K, Ishigami M, Zettl A (2000) Extreme oxygen sensitivity of electronic properties of carbon nanotubes. Science 287:1801–1804

110. Jakubek JZ, Simard B (2004) Two confined phases of argon adsorbed inside open single walled carbon nanotubes. Langmuir 20:5940–5945
111. Andzelm J, Govind N, Maiti A (2006) Nanotube-based gas sensors—role of structural defects. Chem Phys Lett 421:58–62
112. Thomas W, Zhou TC, Kong J, Dai H (2000) Gating individual nanotubes and crosses with scanning probes. Appl Phys Lett 76:2412–2416
113. Chatterjee A. Book chapter (2011): Computer simulation to rationalize structure property correlation of Carbon Nano tube. Springer In a book Title: Carbon and oxide nanostructures, edited by Dr. Noorhana Yahya; Advanced Structured Materials 5, 143–164
114. Chatterjee A (2011) Excited state reactivity index theory: application for small moieties. Int J Quantum Chem 111:3821

Index

MIX
Papier aus verantwortungsvollen Quellen
Paper from responsible sources
FSC® C105338

If you have any concerns about our products,
you can contact us on
ProductSafety@springernature.com

In case Publisher is established outside the EU,
the EU authorized representative is:
Springer Nature Customer Service Center GmbH
Europaplatz 3, 69115 Heidelberg, Germany

Printed by Libri Plureos GmbH
in Hamburg, Germany